REMEMBERING

REMEMBERING

A guide to
New Mexico Cemeteries,
Monuments and Memorials

Text and Photography by
Margaret M. Nava

SANTA FE

Sunstone books may be purchased for educational, business, or sales promotional use. For information please write: Special Markets Department, Sunstone Press, P.O. Box 2321, Santa Fe, New Mexico 87504-2321.

Library of Congress Cataloging-in-Publication Data:

Nava, Margaret M., 1942-
Remembering : a guide to New Mexico cemeteries, monuments and memorials / text and photography by Margaret M. Nava.
p. cm.
Includes bibliographical references.
ISBN 0-86534-486-8 (alk. paper)
1. Cemeteries--New Mexico. 2. Cemeteries--New Mexico--Pictorial works.
3. Monuments--New Mexico. 4. Monuments--New Mexico--Pictorial works.
5. Memorials--New Mexico. 6. Memorials--New Mexico--Pictorial works.
7. New Mexico--Description and travel. 8. New Mexico--History, Local.
9. New Mexico--Pictorial works.
I. Title.
F797.N38 2006
363.7'509789--dc22

2006011407

WWW.SUNSTONEPRESS.COM
SUNSTONE PRESS / POST OFFICE BOX 2321 / SANTA FE, NM 87504-2321 /USA
(505) 988-4418 / ORDERS ONLY (800) 243-5644 / FAX (505) 988-1025

To Bob and Meg
I love you guys!

CONTENTS

PREFACE

Mention the word cemetery and some people conjure up visions of ghosts, goblins, and things that go bump in the night. Others experience intense feelings of loss, despair, and loneliness. I'm not like those people. Show me a cemetery and I immediately start poking around, inspecting the names and dates on the tombstones, looking for unusual epitaphs, and checking out whether or not someone famous is buried there. Is this some sort of morbid fascination? I hope not.

One of the things that first attracted me to New Mexico was the abundance of folk art. Everywhere I looked there were woodcarvings, painted altar screens, pottery sculptures, and tin nichos. I loved it. Most of the pieces were religious in nature and many, if not all, were created by Hispanic artists. Being of Hispanic heritage, I became curious. Why did these people create their beautiful arts and where did they get their inspiration?

A little bit of detective work led me to the High Road about twenty miles northeast of Santa Fe. In this rural region of approximately twelve small villages, Hispanic artists started creating rustic masterpieces more than three hundred years ago The craftsmen—they never considered themselves artists—lived and worked in the piñon covered hills and took their inspiration from a need to stay connected to their roots. Fathers passed their skills on to sons and one generation to the next learned to live in harmony with its surroundings. It was during my trips to the High Road that I started to learn about the Penitentes, a religious group that, among other things, administered justice, nursed the sick, buried the dead, and practiced some very unusual ceremonies.

My first book, *Along the High Road*, tells the story of a people who, tied to the earth, fiercely independent, and staunchly Catholic, settled a hostile land, created a new life for themselves, and became the moral fiber of New Mexico. While reaching that book, I became aware of the beauty of the local camposantos. Most of the tombstones were hand made, some were exceptionally old, and all were given loving care and utmost respect. Flowers adorned the graves, fences surrounded some, and weeds were conspicuous by their absence. Some of the cemeteries near churches were large while others—essentially out in the middle of nowhere—were small.

As I traveled to other areas within the state, I began to notice the other ways people remembered their departed loved ones. Aside from the camposantos, there were cemeteries, monuments, shrines, and memorials of all sorts. Although most of the cemeteries and camposantos were pretty straightforward, there were some rather extraordinary monuments and memorials dedicated to war heroes, animals, bandits, and believe it or not, refrigerators. Each and every cemetery, tombstone, and monument was significant and deserved recognition. So I decided to write a book, not about death and dying, but about the way people remember the people and events that impacted their lives—how and why they remember them—how and why they affected New Mexico.

During my research for this book I learned a great deal about the importance of cemetery preservation, tracing family roots, and making tombstone rubbings. These were areas I knew little or nothing about but since all of this information seemed so intertwined with everything else, I knew it couldn't be left out of this book.

My special thanks go to Dr Virgil N. Kovalenko, Ph.D., USAF (ret) who contacted friends and family in order to answer my questions about the Mormon Battalion Monument at Budaghers; to Chris Faivre, Media/Publications Manager of the Las Cruces Convention and Visitors Bureau who volunteered time and patience in locating the grave of Pat Garret; to Jack Moore, Marketing Director of the New Mexico Museum of Space History who provided the excellent portrait and history of HAM the Astro Chimp; to Dee Turner of Casa San Ysidro in Corrales who told me a story about a flood and rescued caskets; to all the ladies in Carnuel who talked to me about their families buried in the Santo Nino Cemetery; to my cousin Christy Dunlop and her friend Molly Newman who gave up part of their summer vacation to help me find graves; to my dear friends, Lisa and Mario who listened, with interest, every time I started to carry on about a another cemetery or monument; but, most of all, to my brother and his wife who put up with cancelled vacations, sporadic emails, and missed phone calls. Thanks everyone—I couldn't have done this without you!

—M.M. Nava
Rio Rancho, N.M.

INTRODUCTION

This book isn't about death and dying. It's about remembering. Within its pages you will find descriptions and directions to some of New Mexico's unique, sometimes controversial, cemeteries, monuments, and memorials, a beginners guide to tracing your family roots, instructions and cautions on making tombstone rubbings, and information as to how and why to get involved in cemetery preservation. However, in the interest of privacy and due to the extremely sensitive nature of burials, you will not find identification of individual tombstones (except for some rich, famous, or notorious characters that helped shape New Mexico's colorful history) or descriptions and locations of Native American burials.

As in other places, many of New Mexico's cemeteries and monuments have experienced the devastating destruction of vandals and souvenir hunters. Although it shouldn't be necessary, extreme caution is advised when visiting any such site. Please treat all of our cemeteries and monuments with respect. Always get permission to enter private property or family cemeteries; try to not step on any grave; don't let pets run loose in the graveyards or around monuments; don't collect souvenirs—even small rocks; and especially—never desecrate or defile a cemetery, tombstone, monument, or memorial in any way.

So, what is remembering and why do we do it? According to the dictionary, it is the act of bringing someone or something from the past to mind; it is a recollection; a commemoration; a keepsake. To some people, remembering is a way to *re*-member or put something back together, a way of keeping something alive so that we can derive value from it. To others, it is the opposite of forgetting.

According to an ancient belief, a person can die three deaths:
The first death occurs when a person's body ceases to function.
The second death occurs when the body returns to dust.
The third death occurs when a person is forgotten.

Remembering gives us an opportunity to reflect upon our lives, our heritage, and our ancestors. It gives us strength and courage. It helps us understand the meaning of life.

CAMPOSANTOS

New Mexico was a wild territory during the late 1600s and early 1700s. The weather, the terrain, wild animals, hostile Indian attacks, and disease all took their toll on the population. During those early years, the dead were usually buried beneath the dirt floor of a church to protect them against desecration. The more important the difunto (deceased person), the closer he was buried to the altar. Because grave markers weren't used and few permanent records were maintained, the identity and location of many early settlers was forever lost. Following edicts from the Spanish government, church floor burials were eventually prohibited and bodies were buried outside the church in atrios (walled churchyards) or in camposantos (holy fields) usually located some distance from the church.

Camposanto near Chimayo

With the atrios and camposantos came the tradition of graveyard art. Traditionally, grave markers were handmade crosses, most often fashioned from wood, but later constructed from river rock, rough shaped cement, or ordinary plumbing pipe decorated with china doorknobs, curtain rod finials, or small cans. Wrought iron crosses with curvilinear decorations were also popular. Statues, sometimes surrounded by a nicho (grotto like structure) started to appear around the late 1800s when the railroad came through bringing a wider variety of materials and even commercially produced tombstones. Numerous tombstones of mothers and babies lost at childbirth, young children stricken by disease, and men fatally wounded while working or defending their property bear testament to the rigors of living on the frontier.

One of the features that distinguishes camposantos from cemeteries is the use of fences or borders that delineates family plots one from the other and protects the graves from predators, and in recent years, vandals. Like the earliest grave markers, they were originally handmade from wood, wire, iron, or stone but later replaced by professionally forged structures. Most of these structures are simple, unadorned enclosures but some are embellished with paint, marbles, and shards of china or mirrors. The graves of angelitos (deceased children) are especially colorful and they generally have an enclosure at the base of the tombstone filled with the child's favorite toys.

Nichos at San Carlos Camposanto

Another unique characteristic of camposantos is the flowers. Known as coronas (floral offerings), they come in every shape (oval, round, garland, wreath, cross) and almost every grave, tombstone, and fence in a camposanto seems to be covered with at least two or three of them. But these are not ordinary flowers—they are plastic.

Plastic Flowers at Rural Camposant

Coronas were originally made of flores enceradas (crepe paper dipped in wax), but the use of plastic flowers in today's camposantos seems, understandably, widespread. Think about it. The scorching sun, the relentless winds, and especially, the infrequent rain would quickly destroy real plants or flowers. The solution, of course, plastic. Drive past any rural camposanto and you will see a profusion of gaudy plastic flowers made into wreaths and bouquets. Although you might think the use of plastic is a lazy man's way out of taking care of an unhappy chore, eventually, you will agree that it is a loving and thoughtful way of looking after the graves of departed loved ones.

MORADAS

Lacking priests and living in isolated areas, early settlers had to sustain themselves spiritually as well as materially, so they formed a brotherhood that administered justice, supplied sustenance and support to widows and orphans, nursed the sick, and organized village feasts and religious ceremonies.

Calling themselves La Fraternidad de los Hermanos de Nuestro Padre Jesus Nazareno, or more simply Los Hermanos (Penitentes), brotherhood members conducted many of their religious ceremonies in or around modest chapels known as moradas. The moradas were simple structures, typically without windows, very simply decorated: inside with candles and santos (wooden statues) and outside with maderos (heavy wooden crosses).

Replicated Morada at El Rancho de Las Golondrinas

A carved skeletal figure of death, Doña Sebastiana or La Muerte, dressed in black and riding in a carreta de la muerte (cart of death) reminded Penitente members of their mortality and the need to prepare for a good death. During Holy Week, the macabre figure accompanied a procession to the Calavario (Calvary Hill) in a reenactment of Christ's suffering and death.

Doña Sebastiana in Cart at Museo de Santa Fe

Aside from other duties, Los Hermanos members conducted velorios (wakes), funerals, and entierros (burials) for Brotherhood members, families, and neighbors. In her book *Shadows of the Past*, Cleofas M. Jaramillo describes a typical Penitente death observance:

> *When a wake was held for a deceased person, it sometimes lasted two nights, according to the wealth of the deceased or of the relatives. The next morning a homemade pine coffin, containing the remains, was carried in a wagon to the cemetery. The women mourners stayed at the house, lamenting. The resador (reader), the singers, and the crowd of bareheaded people followed the wagon chanting the alabados (religious hymns) in mournful voices.*
>
> *En route to the camposanto . . . a stop or two was made. At each stop, a descanso (wooden cross) was afterwards propped up with a mound of stones to remind the passer-by to pray for the repose of the soul of the dead, whose funeral procession stopped at that place for rest.*

Descanso Along Highway

> *When the deceased belonged to a family whose members belonged to the Penitente Brotherhood, groups of flagellants visited the remains from midnight on, after most of the people had left the wake . . . In the summer the wake was held in the open yard . . . under the silent spell of the night the mournful chants instilling their sad cadence into one's soul.*

When the mourners returned home they placed a picture of their deceased family member on their ofrenda (home altar), surrounded it with the departed one's favorite foods or pastimes, lit a candle, and included the name of the departed one in their nightly prayers.

In recent years, wakes have been reduced to one night or less, hearses have replaced funeral carts, descansos are now used to memorialize the site of a fatal road accident, the number of home altars has diminished, and membership in the Brotherhood has declined largely due to changes in values, attitudes and outside scrutiny. Are the old ways dying? Maybe. Maybe not. That's one of the questions this book hopes to answer.

CEMETERIES

Most of the graves in camposantos were shallow and the majority of corpses were not embalmed. Flooding, animals, and grave robbers often disturbed graves, dislodging those buried within and causing health hazards. For early settlers this was more of an inconvenience than a problem but as New Mexico's population grew and became more urbanized, the need for larger, more hygienic cemeteries became apparent. Garden-like cemeteries embellished with flowers, trees, and grand pathways became commonplace. Mausoleums, columbariums, and large, manufactured monuments replaced homemade tombstones and grave markers and the popularity of camposantos declined.

Unlike camposantos where burials appear random and disorganized, most of the graves in a contemporary cemetery are lined up in neat, almost impersonal, rows of tombstones all facing the same direction. A perfect example is the National Cemetery in Santa Fe. At this cemetery, each grave is distinctly marked and its location is recorded on a user-friendly computer that prints gravesite maps. The roads are clean, wide, and clearly identified, and it is fairly easy for family members to find the neatly groomed grave of their departed relatives.

Santa Fe National Cemetery

Such is not the case at all cemeteries, however. In fact, right next door at the Rosario Camposanto, things are quite a bit different. Graves appear to have been placed wherever there was room. The roads, where there are any, are mostly dirt tracks leading to dead ends. Few, if any, are marked. Over the years some of the tombstones at Rosario were removed, vandalized, or stolen.

Broken Tombstone in Rosario Cemetery

Why do things like this happen? The problem is funding. The National Cemetery is nationally funded while maintenance of Rosario relies on contributions from families, some of whom no longer exist, or worse, no longer care about the graves. Sad, but true, this disturbing fact makes a strong case for the value and importance of cemetery preservation that is addressed in chapter eight of this book.

MONUMENTS AND MEMORIALS

New Mexico's colorful and exciting past is recognized and honored in its many monuments. Some, like Capulin Volcano, El Morro, and White Sands relate to prehistoric events while others, like Aztec Ruins, Bandelier, Petroglyph, and the Gila Cliff Dwellings recall the history of early man. Many pay tribute to the men and women who helped shape the future of the state. There are monuments depicting early explorers and missionaries, monuments and markers dedicated to the Mormon Battalion, Civil War, WWI and WWII monuments, Veterans walls, parks and highways and, of course, Vietnam, Desert Storm, and Iraq memorials. There are even some monuments that commemorate famous and infamous people who once lived here. Some were bandits and lawmen; others were writers, artists, or martyrs. Each contributed an important chapter to New Mexico's history and each is remembered with respect if not admiration.

Vietnam Memorial at Angel Fire

Turning to things less serious, there are also some rather unusual memorials in New Mexico. For example, our old friend Smokey Bear looms large in Capitan; Ham the Astro Chimp holds a place of honor at the Museum of Space History in Alamogordo; and Fridgehenge—an outright mockery of unrestrained consumerism—sits high atop a Santa Fe hill. Seem strange to pay homage to an animal or a bunch of refrigerators? Once you see these places, you may change your mind.

And what about the festivals? Aside from Indian Dances, wine festivals, balloon fiestas, and arts and crafts shows, there's the Kings' Day celebrations in January, the Cowboy Roundup in February, the Lady on the Mountain Folk Festival in March, the Trinity Site tours in April and October, the Memorial Day celebrations in May, the Old Fort Sumner Days in June, the Smokey Bear Stampede in July, the Inter-Tribal Indian Ceremonial in August, the burning of Zozobra in September, the Day of the Dead in October, the Veterans' Day ceremonies in November, and the Indian, Hispanic, and Anglo Christmas celebrations in December. Aren't festivals just another way of remembering?

Several centuries back, three cultures came together to form the Land of Enchantment, as we now know it. Each culture had its own way of life and each had rituals and ceremonies that celebrated its history, achievements, folklore, and beliefs. Since those early days, historic places have been preserved, monuments have been built, many of the old rituals and ceremonies have been revived, and new memorials and celebrations have been created. Regardless of origin, each historic place, ritual, memorial, and ceremony is unique and each represents an important chapter in New Mexico's history. However, since it would take several volumes to list and describe every one of them, only a few were chosen to be included in this book.

1

NORTHERN NEW MEXICO

VELARDE SHRINES

Location: 11 miles NE of Española off NM 68 in Velarde
Established: Church 1817—Chapel 1980s
Features: Bulto of Jesus the Divine Mercy
Contact: San Antonio Church in Peñasco / (505) 587-2111

The fertile valley of the Rio Grande is well known for its apples, chiles, and wine but few people are aware of the treasures that lie all but hidden along the banks of this great river in the village of Velarde.

Iglesia de la Virgen de Guadalupe (The Church of the Virgin of Guadalupe), was built in 1817 or about the same time Spanish settlers in New Mexico were trying to win their independence from Spain. Like many other churches of its time, the mission church was built with four-foot thick walls, a territorial-style pitched metal roof, and a large buttress at the rear. The church was dedicated to the Blessed Virgin who appeared to Juan Diego, a poor shepherd who carried rose petals to his bishop as proof of the Blessed Lady's appearance. For years, the church stood bereft of any decoration other than a small wooden altar and a few stained-glass windows. Then, something amazing happened.

Iglesia de la Virgen de Guadalupe

When weaver Zoriada Ortega learned she had cancer, she prayed to Our Lady of Guadalupe and made a promesa (promise) that if she were healed she would have a capilla (small family chapel) built much in the same way as Juan Diego did in 16th century Mexico. The Virgin of Guadalupe was always a great source of empowerment for Zoriada so it came as no surprise that she was cured of her cancer without having to endure chemotherapy. In order to help Zoriada carry out her promesa, her husband Eulogio hired three men to lay a foundation for the family chapel but when the men discovered that the chapel was to be dedicated to la Virgen de Guadalupe, they completed the entire structure.

Eulogio was a santero and, although he had been working on bultos (three dimensional wooden santos) for several other churches, he set to work carving a statue of Our Lady of Guadalupe for his wife's capilla. Seeing all the work her husband was putting into the statue, Zoriada decided to help by painting la Virgen's blue cloak, rose-colored dress, and dark skin. After the bulto of Our Lady of Guadalupe was complete, Zoriada and Eulogio worked together on several other santos. One in particular traces its history to Poland.

In 1931, Jesus appeared to a Polish nun, Sister Maria Faustina, and told her to paint His image and sign it, "Jesus I trust in you." Although the nun wasn't happy with her painting, the Lord told her that if people around the world venerated His image, His Divine Mercy would save them. The nun prayed before the statue every day and wrote more than 600 pages of daily devotions that eventually became known as the Revelations of Divine Mercy. Sister Faustina died in 1938 but her devotions were translated into several different languages. Some of the translations were confusing or erroneous so the Vatican banned any further translations and all devotions to the Divine Mercy. Several years later, Cardinal Wojtyla, the then archbishop of Krakow, initiated a study on the writings of the Polish nun and made recommendations to have the Vatican lift the previously issued ban. In 1978, Pope Paul VI lifted the ban and permitted the devotions to resume. Pope John Paul II (the former Cardinal Wojtyla) beatified Sister Maria Faustina Kowalska on April 18, 1993 and canonized her on April

30, 2000, the first Sunday after Easter, a day that would thenceforth be celebrated as Divine Mercy Sunday.

Eulogio and Zoriada read Sister Faustina's biography and began praying the daily devotions. They were so moved by the nun's words and her devotion to Jesus that they decided to create a bulto of Jesus the Divine Mercy so that others could experience the Lord's heavenly benevolence. Eulogio spent al-most five months carving a 39-inch tall image of Jesus with His nail-pierced hands raised in supplication and twin rays of water and blood radiating from His heart. Water cleanses the soul and blood gives it life. Zoriada spent an additional two months painting the statue.

Jesus Divine Mercy Santo

When the parish priest saw the bulto, he was so overwhelmed with the statue's simplicity and beauty that he asked if it could be moved from the capilla to the church. On April 6, 2002, the bulto of Jesus the Divine Mercy was placed in an empty nicho in the old church where, along with la Virgen de Guadalupe and other santos created by the Ortegas, it has gained the status of a shrine.

Sometimes, when you make the time and effort to seek out some of New Mexico's out-of-the-way monuments and shrines, the rewards are amazing.

JUAN DE OÑATE MONUMENT

Location: 8 miles NE of Española, on NM 68 in Alcade
Established: 1994
Features: Bronze Statue and Visitor's Center
Contact: Visitor's Center / (505) 852-4639

Depending on whom you talk to, Juan de Oñate was either one of New Mexico's worst criminals or one of its greatest pioneers.

It was the winter of 1598. After three years of fighting his way through a gaggle of political red tape, Don Juan de Oñate, aristocrat, mine-owner, soldier, and would-be-adventurer, crossed the Rio del Norte with his son, three friars, five missionaries, two lay brothers, 100 Indians, and almost 200 men, women, and children in 83 covered wagons and headed north toward the final frontier of New Spain in search of land, riches, and souls. It took six months to cross the Chihuahua Desert but by July, the expedition, half-starved and miserable, entered the Pueblo village of Okeh Owinge in northern New Mexico, drove out the Indians, moved livestock in, and renamed the village San Juan de los Caballeros (Saint John of the Gentlemen). Shortly thereafter, Oñate moved the settlers and all the livestock to a new location across the Rio Grande, known as San Gabriel del Yunque. The new location became the territory's first capital with Oñate serving as its governor.

Juan de Oñate Statue

Almost immediately, there were problems. Disappointed at not finding "bars of silver on the ground . . . and resentful because (Oñate) did not allow them to abuse the natives," some of the colonists attempted mutiny. Oñate handled the situation by beheading two of the agitators. The worst was yet to come.

In the fall of that same year when Oñate abandoned the settlement in the hopes of finding the South Sea, which he erroneously believed bordered New Mexico, he sent word for his nephew, Juan de Zaldívar, to form a scouting party and meet up with him somewhere near Zuni. Along the way, Zaldívar and his men camped at the base of a 367-foot high sandstone mesa upon which the Pueblo village of Acoma was perched. While looking for food and supplies, the campers climbed to the top of the mesa and either surprised the natives that lived there or fell into their well-laid trap. The Indians attacked, slaughtered 13 Spaniards, including the governor's nephew, and threw their victims' bodies over the steep walls of the mesa.

Oñate deliberated for several weeks before taking action but by January of 1599, he decided what to do. Seventy men (not including himself) armed with muskets, cannons, and swords and dressed in heavy armor, stormed the Acoma mesa. A three-day battle ensued, culminating in the death of 800 natives and the capture of hundreds of Acoma males who, following a ad hoc Spanish trial, were sentenced to 20 years in prison, had a foot amputated, or were forced into slavery.

Overwhelmed by Oñate's tyrannical behavior, the colonists wrote letters to the Spanish government complaining that not only was Oñate cruel to the natives, he ruled the colony with an iron fist and was often absent from the settlement leaving the colonists to their own defenses.

Things began to happen quickly. The Spanish Viceroy reprimanded Oñate for his hostility toward the natives and his alleged dereliction of duty. Oñate responded by bringing in more soldiers and friars and then going out, once again, to search for the elusive South Sea. Colonists deserted the settlement. Friars abandoned their church. Neither gold nor silver was ever found. In1608, Pedro de Peralta was sent in to succeed

Oñate as governor and by 1610, the capital was moved to Santa Fe.

In Oñate's defense, it must be pointed out that he was a businessman and a gentleman, not an explorer. When he volunteered to personally finance an expedition into the Northern Frontier, he did so without knowing what faced him. Although there were times when supplies ran short and the terrain seemed impassable, Oñate pushed his expedition forward through the desert, across the river, and over the mountains. In so doing, he blazed a trail from San Bartolomé, Mexico to San Juan de los Caballeros in New Mexico. At the end of the trail, Oñate founded the first Spanish community in the northern frontier (nine years before the founding of Jamestown and 22 years before the pilgrims landed at Plymouth Rock), he built the first Christian church to be constructed in what is now the United States, and he established the frontier's first capital.

El Camino al Norte (the Road North) that Oñate and his group followed became an integral part of El Camino Real (the Royal Road) that served as passage for settlers, livestock, Christianity, and agriculture and altered the face of the American Southwest. True, there were more expeditions after Oñate's but had he not blazed the trail, those expeditions might have ventured elsewhere.

Biographer Marc Simmons called Oñate "the godfather of the Franciscan missionary program on the Northern Frontier," and states that Oñate "inaugurated mining and the process of ores, launched the livestock industry, and opened the first major road in the Southwest—El Camino Real." In short, as Simmons' title suggests, Oñate was "The Last Conquistador."

In 1994, a 12-foot tall bronze statue of Don Juan de Oñate on horseback was erected in the village of Alcade, not far from where Oñate's breakthrough expedition came to an end at the Pueblo of Okeh Owinge. Needless to say, the statue caused quite a commotion. Hispanics living in the area thought the statue would be an appropriate addition to the celebration of the upcoming Quattro Centennial of the founding of the first Spanish colony in the Northern Frontier. Native Americas disagreed. Intense arguments broke out on both sides but nothing much happened until just before the celebration was to begin.

According to a 1998 news article:

Española, N.M.—One moonless night in early January, just as Hispanic New Mexicans were starting to celebrate the 400th anniversary of the first Spanish settlement in the American West, an American Indian commando group stealthily approached a bronze statue here of the conquistador, Don Juan de Oñate. With an electric saw, the group slowly severed his right foot—boot, stirrup, star-shaped spur and all. "We took the liberty of removing Oñate's right foot on behalf of our brothers and sisters of Acoma Pueblo," read a statement sent by the group . . . "We see no glory in celebrating Oñate's fourth centennial and we do not want our faces rubbed in it."

Onate Boot

The foot and boot were welded back in place but a few years later, tempers flared again when Albuquerque city officials announced a similar statue would be erected in one of that city's parks. Hispanics argued that Oñate was part of their history and that he should be honored. Native Americans also believed that Oñate was part of their history—a part they would rather forget. After several months, numerous meetings and lots of name-calling, the arguments could not be resolved and the city was forced to abandon the project.

So was Oñate a villain or a hero? I guess it all depends on your point of view.

ESPAÑOLA VETERAN'S MEMORIAL WALL

Location: Across from the Old Mission near Paseo de Onate & Bond Streets in Española
Established: 2003
Features: Memorial to All Veterans
Contact: Española Valley Chamber of Commerce
(505) 753-2831

On November 11, 1918, representatives of France, Germany, and Britain met in a train car outside the French town of Rethondes and signed the armistice, or truce, that ended World War I. The fighting stopped at 11AM on the 11th day of the 11th month. One year later, President Woodrow Wilson asked workers, students, and homemakers to observe two moments of silence to honor the men who died in that war. From then on, the silence became an unofficial tradition observed each year throughout most of the United States. On June 4, 1926, Congress enacted a resolution asking the President of the United States to issue a proclamation calling for official observance of this significant day. Act 52 Stat.351 was approved on May 13, 1938 declaring November 11th a legal holiday "dedicated to the cause of world peace and to be hereafter celebrated and known as Armistice Day."

Armistice Day was primarily a day set aside to honor veterans of World War I but in 1954, after World War II necessitated the largest mobilization of soldiers, sailors, marines, and airmen in the Nation's history and after American forces had courageously fought in Korea, the 83rd Congress amended the Act of 1938 by striking out the word "Armistice" and inserting the word "Veterans." So it was that November 11th became a day to honor American veterans of all wars.

Years later, following a Veteran's Day ceremony honoring Korean War veterans, Española City officials began talking about building a memorial to honor all of Northern New Mexico's veterans from World War I through the Gulf Wars. At first, residents of this historic village weren't very optimistic but with the help of Senator Pete Domenici, $750,000 in federal and state money was raised for the project.

The wall, designed to allow space for the names of future veterans, was dedicated on November 11, 2003 and lists the names of more than 2000 veterans from Española, Pojoaque, Nambe, San Juan, San Ildefonso, Santa Cruz, Chimayo, Truchas, Peñasco, and Picuris who served in the U.S. Armed Forces.

Española Veteran's Memorial Wall

In his speech commemorating Veterans Day 2003, Domenici addressed the Senate and urged Americans to take a step back and reflect on the sacrifices made by our veterans, especially those from New Mexico.

Today, we have so many millions of American men and women who... went off to serve their country. They faced times of great fear and great concern. They experienced times where there was heroism all around them. I just want to say to all our veterans in all our wars, thank you on behalf of the people of (our) state for what (you) have done to preserve the greatness of this country and the concept of freedom and liberty.

This wall in Española will serve two important purposes. First, it is a symbol of appreciation to veterans of the past. Second, it is a strong message of support to our veterans of the future. To me, that kind of appreciation and support is a civic duty and I applaud Española for the vision and the will to build this veterans memorial. It is my sincere hope that it will serve to inspire many for years to come.

The Española Veteran's Memorial Wall is near the old town plaza, across from the old Mission and Convento, a replica of Don Juan de Oñate's first church built in 1598. On the wall are the words "A Community Gratefully Remembers." A Liberty garden surrounded by roses and a bubbling garden fountain give the memorial a park-like appearance. Low Riders, those brightly painted automobiles with small tires and great hydraulics, cruise the streets that surround the wall. Some of the cars' drivers are young; others are not. Maybe some of them served in a war or are getting ready to do so. Will they forget those who have gone before?

Not likely!

THE VILLAGE OF HERNANDEZ

Location: 5 miles northwest of Española on US 84-285
Established: Early 1800s
Features: Abandoned church, Ansel Adams photo site
Contact: None

As the Rio Chama winds its way down through the pastel cliffs and red canyons between El Vado and Española, it passes through the community of Hernandez. The town, probably settled sometime after the Reconquest of 1692, was originally named San Jose de Chama but later took the name of one of its original settlers.

For as far back as anyone can remember, life in this small village was peaceful and quiet. In the early years, people who lived here took their livelihood from farming and ranching. In 1820, they built a small church and dedicated it to the village's patron saint, San Jose, spouse of the Virgin and foster-father of Jesus. Children were born; families prospered; those who died were buried in the camposanto behind the church. Some time later, writers, spiritualists, and artists discovered the ambiance of the area was conducive to their ethereal pursuits. One of those artists was Ansel Adams.

A long time friend of Georgia O'Keefe, Adams often took advantage of his visits to Ghost Ranch to photograph the surrounding countryside. On one such occasion in 1941, while driving back to Santa Fe after a particularly frustrating day of being unable to capture any suitable photographic images, Adams caught a glimpse of the three-quarter moon rising above the San Jose de Chama church and camposanto. It was that critical time of day when there is just enough sunlight to illuminate landscape features but not enough to obscure the moon. Adams couldn't believe his luck. There it was, the picture he had spent all day looking for. But he had to hurry. Time was running out and within minutes sunset would change the idyllic scene.

I almost ditched the car and rushed to set up my 8 X 10" camera. I was yelling to my companions to bring me things from the car. I had a clear visualization of the image I wanted but I couldn't find my light meter. The situation was desperate. The low sun was trailing the edge of the clouds in the west and shadow would soon dim the white crosses.

Even though Adams attempted to take a second shot, the moment was gone. Nightfall overtook the valley, the church fell into deep shadows, and the white crosses of the camposanto disappeared into the dark.

Rushing back to his studio, Adams began work on the negative. Using a variety of methods, he experimented with several different chemicals, development times, and papers. With the negative in the enlarger, he increased the light hitting certain areas to make the sky blacker and the clouds less bright so that the moon would stand out more. After many attempts, he achieved his goal: a dramatic image that he called "Moonrise—Hernandez, New Mexico."

Adams went on to photograph other landscapes and other images but throughout his life, "Moonrise" remained his favorite because it "combined serendipity and immediate technical recall." When he took the photo, Adams said he felt "an almost prophetic sense of satisfaction."

Adams, one of the greatest photographic artists of the twentieth-century, was often criticized for not including humans in his photographs and for representing an idealized wilderness that no longer exists. However, his impression of the village of Hernandez and its small church could not have been more expressive had it contained two or three people walking around or tending to daily chores. As it is, the photograph conveys the peacefulness of the village, the sanctity of the church, and the promise of resurrection as evidenced by the moon rising above the camposanto.

These days, it's hard to tell if the old church and camposanto still exist. A new church and cemetery were built in 1972 but people at the new church seem reluctant to answer any questions. Manufactured and mobile homes have sprung up around the old church's reported location and there are no signs directing sightseers through the neighborhood.

San Jose Cemetery at Hernandez

In June of 2000, many of the old church's santos and Stations of the Cross were stolen. Los Hermanos assumed the care of the church and a group known as the San José de Chama Restoration Committee was formed to restore the church to its original condition. Volunteers apply mud to the old adobe walls, repair roof leaks, and rebuild decaying buttress walls. Maybe, one day, the old church will once again be opened so that people may take comfort in he sanctity and serenity of this landmark shrine.

Until that day, the moon will still rise over Hernandez—even if the only one who sees it is a coyote.

CERRO PEDERNAL

Location: 18 miles northwest of Española off US 84-285 near Abiquiu
Established: Prehistoric
Features: Pedernal Peak and Georgia O'Keefe House and Studio
Contact: Santa Fe National Forest / (505) 438-7840
O'Keefe Foundation / (505) 685-4539

When Georgia O'Keefe first visited New Mexico in 1929, she knew she'd be back. The dry, thin air made colors more vivid and, somehow, allowed her to see farther than she ever had. This was the perfect place to paint. In 1934, she began spending summers at the legendary Ghost Ranch, driving around in her Model A Ford and painting evocative fragments of views she beheld. She was so fascinated by the ranch (known to locals as El Rancho de los Brujos—the Ranch of the Witches) and its surroundings that she bought the cottage she always stayed in, opening it up to friends and fellow artists who were able to find their way

Cerro Pedernal

through the daunting mountains. Three years after the death of her husband, Alfred Stieglitz, O'Keefe finally made New Mexico her permanent home. But if it was dry air and vivid colors that first attracted Georgia, it was Cerro Pedernal that eventually captivated her.

Cerro Pedernal is a 9,862' anvil shaped bluff that dominates the northeastern end of the Jemez Mountain range. Thousands of years before the dawn of the Pueblo culture, ancient huntsmen dug in the 15-foot-thick layer of chalcedony (agate) that occurs below the mountain's flat-top to obtain flint for their arrowheads and spear points. Centuries later, explorers discovered the flint and named the peak Pedernal which, in Spanish, means flint. Hoping to find riches, colonists that followed the explorers carved out mines and built tunnels on the mountain's slopes. When the expected riches weren't found, the mines and tunnels were abandoned and the colonists moved on. But this was where O'Keefe wanted to be and this was where she built her home.

In December of 1945, O'Keefe bought a three-acre, dilapidated farm in Abiquiu and began the arduous task of turning it into her new home and studio. The roof of the main house was caved in, the out buildings were windowless, the trees in the orchard were gnarled and fruitless, and chickens freely roamed in and out of the ruins as if they owned them. When Georgia wasn't busy renovating decaying buildings, trying to obtain construction permits, or looking for building materials, she took pleasure by painting bold but sensitive representations of her new surroundings. She painted the river, the cliffs, the skies, wildflowers, animal skulls, churches, doors, and, of course, Cerro Pedernal.

El Cerro became O'Keefe's touchstone. Like windows in time, she captured its essence in the early morning, the late evening, throughout the seasons and in every mood. She saw the peak shrouded by clouds, scorched by the sun, tortured by storms, and kissed by rain. Its interwoven layers of maroon and green shale became familiar old friends and Georgia included the mountain in many of her landscapes. She was fond of saying, "Pedernal is my private mountain. God told me, if I painted it often enough, I could have it." And hers it became.

As the years ticked away, friends tried to convince Georgia to move into town. For the longest while, she resisted all of their pleas but when her health and eyesight failed and she could no longer find helpers who would stay with her, O'Keefe relented and moved to Santa Fe where she died, at the age of ninety-eight, on March 6, 1986.

In his book, *O'Keefe: The Life of an American Legend,* biographer Jeffrey Hogrefe relates what happened next:

> *Both the New York Times and the Washington Post reported her death on the front page of their March 7 editions, but the passing of an American legend was otherwise a quiet and personal affair. There was no memorial. No one spoke or read from inspirational passages. There was no singing. Carrying out her final instructions, Hamilton (O'Keefe's companion and caretaker) walked to the top of the Pedernal bearing an urn containing her cremated remains. He tossed the ashes into a strong wind.*

Georgia Totto O'Keefe died as she had lived—quietly, privately, and without a lot of fanfare. She was the best known and most celebrated woman artist of the twentieth-century but, possibly, the least understood. She was a woman before her time. She was unconventional, controversial, audacious, and aggressive but she believed in life and lived it to the fullest. As she once told a friend, "Kick your heels in the air." Maybe that's the way she wanted to be remembered—kicking her heels up on the top of Cerro Pedernal.

2

THE HIGH ROAD AND TAOS AREAS

LA IGLESIA DE SANTA CRUZ DE LA CAÑADA
(The Church of the Holy Cross of the Canyon)

Location: 2 miles east of Espanola on Santa Cruz Road (State Road 76)
Established: 1730
Features: Camposantos and Santo Entierro
Contact: Church Office / (505) 753-3345

In the early 1600s, settlers built a church when they and Don Juan de Oñate arrived in what was later called Santa Cruz (Holy Cross). The church, however, suffered extensive damage during the Pueblo Revolt of 1680 and the twelve years following. Governor Cervasio Cruzat y Gongora who inspected it in 1732, declared the church "beyond repair and in danger of collapsing." In 1733, the settlers of La Villa Nueva de Santa Cruz de la Cañada obtained official permission to build a new church on land donated by a wealthy widow. It took fifteen years to finish the structure but by 1748, La Iglesia de Santa Cruz de la Cañada was complete.

La Iglesia de Santa Cruz de la Cañada

In sharp contrast to its elaborate 15th century predecessors, La Iglesia de Santa Cruz was a simple mud and adobe church, built in the traditional cruciform shape, with a flat roof, two bell towers, adobe walls that were more than 32 inches thick, and a dirt floor under which many early settlers were buried. Unfortunately, like many churches of the time, La Iglesia lacked the religious paintings and sculptures essential in the process of conducting religious services. The colonists believed their services were incomplete without these symbols and began creating rustic santos (religious symbols) that represented Christ, the Holy Family, saints, and angels. The men who created the santos, many of whom were Franciscan friars, were called santeros.

Fray Andrés Garcia, a Franciscan painter and sculptor, was assigned to Santa Cruz from 1765 through 1768. One of the few 18th century New Mexico santeros ever identified, Garcia produced a decorative, though traditional, altar rail, altar screen and several carved images. What little we know about the friar is that he used native materials, experimented with proportion and detail, employed color to simulate depth and perspective, and applied gesso to all of his panels. One of his works is the Santo Entierro, a life-sized carving of the crucified Christ.

With its articulated arms and legs and realistic nail holes covered in blood, the Santo Entierro rests in an open-sided sarcophagus located in a small alcove in the nave of the church at Santa Cruz. A crown of thorns encircles His head and a white cloth is wrapped around His body. Most likely created for use by the Penitentes, the statue was used during La Semana Santa (Holy Week) to emulate Christ's capture, scourging, crucifixion, and burial.

The Holy Week ceremonies were the most important ceremonies performed by the Penitentes. Beginning with Los Matines de las Tiniebas (afternoon of darkness) vesper services on Spy Wednesday, Brotherhood members carried the Santo Entierro in several processions to the Calvario (hill) where the crucifixion would be reenacted. On Wednesday and Thursday, the Santo was clothed in a red cape and referred to as Jesus Nazareno. On Good Friday, the cape was removed to reveal the brutality of the Crucifixion as well as the somberness of the ceremony that followed. In a 1952 article in *New Mexico Quarterly*, Florence Hawley Ellis describes one version of that ceremony:

> *On Good Friday the red cape was removed and following an arduous procession up the calvario, the santo was "hung upon the cross, the nails thrust through holes provided in hands and feet. Later, after the scene of the crucifixion, He was taken down from the cross, replaced in the coffin, and returned to the church."*

During the 18th and early 19th century, Spain maintained rule over the New Mexico frontier. Throughout its years of domination, the Royal Government issued decrees and proclamations aimed at gaining and exercising authority over the settlers under its rule. One of the decrees concerned church floor burials.

The practice of placing burials beneath a church floor was common in New Mexico. Not only did church floor gravesites provide protection against desecration by Indian raiders, settlers found it easier to dig a grave inside a church than out of doors when the ground was frozen. In 1798, following a devastating outbreak of smallpox, Spain issued a cedula real (royal decree) forbidding burials in churches in the interest of public health. A similar ruling followed in 1819. Some people reluctantly complied with the edicts and began burying their dead in atrios (walled church yards) but others objected, probably because of the prestige and honor conferred through burial inside the church.

The Mexican government took it a step further when it acquired control of the frontier in the mid-1800s. Now, instead of burying loved ones beneath the church floors or in a courtyards outside the church, colonists were directed to cease and remove such burials to locations away from the churches. Even though this came as quite a blow to the families living near Santa Cruz, by the mid-1800s the dirt floor of la Iglesia de Santa Cruz de la Cañada was covered with wood, burials in the atrio were, for the most part, discontinued, and a camposanto (holy field) was set up on the west side of the road, away from the church.

Camposanto at Santa Cruz

Church records indicate that the first camposanto, known as El Camposanto Viejo, was used extensively until 1942 when Don Roberto Quintana and his wife, Adelaida, donated one acre of land on the east side of the road for Camposanto Nuevo. In 1946, the Quintanas donated another three acres so that this second camposanto could be expanded. Then, in 1967, Don Jose Ygnacio Madrid and his wife, Donelia Lujan Madrid, donated five acres of land located south of Camposanto Nuevo. The donation was the wish of Don Eligio Madrid, Jose Ygnacio's father, who during his life expressed a wish that some of the family's land be donated to La Iglesia de Santa Cruz. The land is now utilized as a parking area for people attending funerals. Possibly, sometime in the future, it will become yet another camposanto.

Although there has been considerable restoration to La Iglesia de Santa Cruz de la Cañada, the camposantos that surround it stand as enduring memories of the piety, strength, self-sufficiency, and courage of the people who survived isolation, Indian attacks, and censure in order to create new lives for themselves and their families.

SANTUARIO DE CHIMAYO

Location: Off State Road 76 in Chimayo
Established: 1816
Features: Healing soil, National Historic Landmark
Contact: 505-351-4889

If only one place could be chosen to exemplify the true nature of New Mexico, it would have to be the Santuario de Chimayo. One of New Mexico's most famous shrines, the Santuario has always been a place of tranquility and spiritual renewal and, to this day, it remains a place of great significance and memories.

Santuario de Chimayo

The first people to occupy the community now known as Chimayo were Tewa Indians. One of their ancient legends recounts a battle between twin war gods and an evil monster that devoured children. The battle lasted several days, but the war gods finally won out. As the fiendish creature died, the earth began to smoke, and flames rose up from deep within a pond the Indians called Tsimajopokwi. Because good had overcome evil in this place, the Indians considered it a source of great healing and used its waters both to cure and ward off various illnesses. Eventually the water evaporated and turned to mud, but the healing practices continued. Even when the mud finally turned to dust, the Indians traveled great distances just to rub some of the health-giving soil on their ailing bodies.

Hundreds of years later, Spanish colonists moved into the area. Located on the eastern boundary of the Northern Frontier, the new settlement was regarded as the outermost edge of civilization. Its mountains and hills were treacherous and isolated, the weather was characterized by heavy rains and cold winds, and the Indians whispered about formidable giants living in the numerous caves that dotted the landscape. What better place to confine deserters, rebels, and thieves? Although no official records exist, many historians believe settlers incarcerated such troublemakers, often forcing them into servitude and military conscription, in Tsimayoh until the Pueblo Revolt of 1680 when they, and their prisoners, were run out of the valley.

Twelve years after the Revolt, several groups of settlers moved back into the area. Threatened by continued hostile Indian attack and fearing for their lives, the colonists built their homes around a defensible central plaza, the Plaza de San Buenaventura, now known as Plaza del Cerro. With only three entranceways into the fortified compound, the homes faced inward toward the main square. Families and animals were sheltered within the compound's walls and a life-giving acequia (ditch) ran through the center of the plaza. Just down the road, in an area probably used as pastureland, a separate settlement developed. Known as El Potrero, this settlement was the home of Don Bernardo Abeyta, an honored member of the Penitentes who, as legend records, discovered a mysterious crucifix near his home.

On a Good Friday in the early 1800s, Don Bernardo left his home in El Potrero to perform the society's traditional penances near the ancient Indian pond. On the way, he was startled by a sudden burst of light. Though frightened at first, Don Bernardo realized that the light was coming from the ground just a few feet ahead. Thinking it might be lost gold from the fabled Seven Cities of Cibola, he fell to his knees and started digging with his bare hands. What he found was a six-foot-tall, dusky-green crucifix bearing a dark-skinned corpus. Abeyta immediately recognized the crucifix as that of the Black Christ, Our Lord of Esquipulas, which had first been brought to the area by a Guatemalan priest more than a century earlier. During the Pueblo Revolt, the priest had been killed and his body buried along with the cross.

Overwhelmed by the discovery, Don Bernardo rubbed his dirt-stained hands across his face and body. Later he would tell neighbors that the soil in which the crucifix had been buried cured him of an ongoing illness; but for the time being, his only concern was to notify the local priest, Father Sebastian Alvarez.

Father Alvarez took the crucifix back to his church in Santa Cruz and installed it in a place of honor on the main altar. The next morning when he went into the church to say Mass, the crucifix was gone. Fearing the worst, he set out to tell Don Bernardo that the crucifix had been stolen. But before reaching Abeyta's hacienda, he noticed a group of people gathered around the place where the crucifix was first found.

"What is going on?" he asked. The people stood back and pointed to the ground. The crucifix had returned.

Once again, Father Alvarez transported the crucifix to the church in Santa Cruz, and once again, it reappeared where it had been found. After yet a third failed attempt, it was agreed that the crucifix belonged in Potrero.

Abeyta had already built a hermita (small shelter) over the holy ground, leaving a pozito (hole) in the floor so that others could touch the tierra bendita (holy earth) and perhaps be healed as he had been. With permission from the church in Santa Cruz, he began construction of a chapel to house the crucifix.

Two rooms, one to the left and one to the right, formed the narthex (vestibule) of the chapel. Within the nave or main body of the church, were bultos (carvings) and retablos (paintings) depicting saints, liturgical feasts, and members of the Holy Family. Behind the main altar, the treasured crucifix stood surrounded by a large reredo (altar screen) with paintings depicting the Jerusalem Cross, the Franciscan emblem, and a cross with a lance, rod, heart, and the four wounds of Christ. To the left of the altar was the original hermita with the pozito for the tierra bendita, and a prayer room. It took several years to complete construction, but by 1816, El Santuario de Nuestro Senor de Esquipulas was finished. Word quickly spread about Abeyta's cure and before long pilgrims were making trips to the chapel searching for miracles of their own.

El Santuario remained a privately owned chapel until 1929 when, after much neglect and financial difficulties, title to the property was transferred to the Archdiocese of Santa Fe. Under the careful direction of the priests of the Congregation of Sons of the Holy Family, El Santuario was restored. As more and more people came to worship in front of the

Rosaries and Canes in Prayer Room

large crucifix and to seek the healing power of the earth, more and more accounts of miraculous cures were added to that of Abeyta's. Since that time, El Santuario has been designated a National Historic Landmark and is frequently referred to as the "Lourdes of America" because of its similarity to the famous shrine in France.

It has been said that the discarded canes, braces, wheelchairs, and messages of thanksgiving that hang from the adobe walls in the prayer room are proof of the miracles of Chimayo. Still, while many people have left their crutches and walked away cured, the Catholic Church has never sought to officially confirm or deny any of the miracles. It has also been said that the dirt in the pozito replenishes itself. Yet, it is common knowledge that the dirt is brought in from surrounding hillsides and, though blessed by a priest, has no special power in and of itself. Even so, in this day and age, it's comforting to know that there are people who still believe that miracles can, and do, happen.

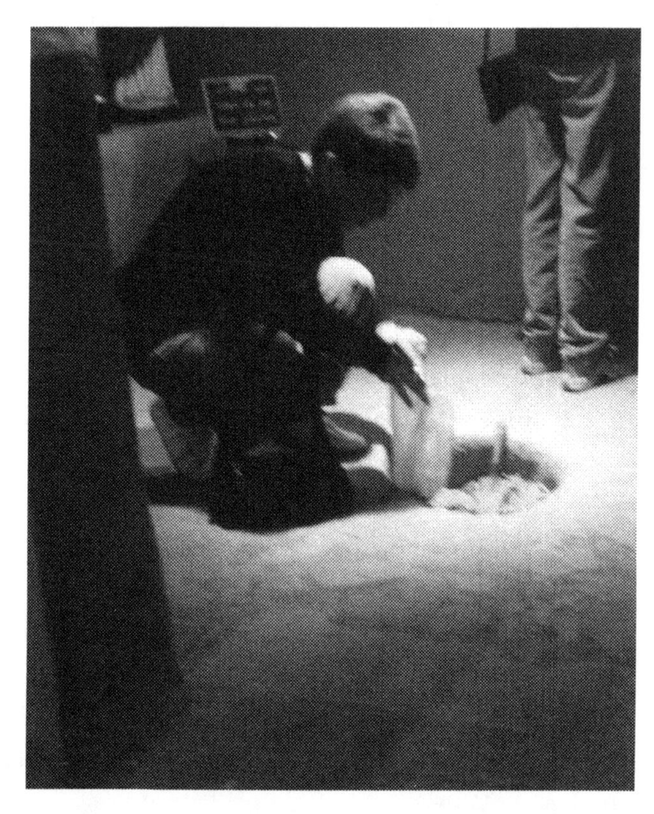

Pozito at Santuario de Chimayo

LA CAPILLA DEL SANTO NIÑO DE ATOCHA
(The Chapel of the Holy Child of Atocha)

Location: Next to Santuario de Chimayo off State Road 76
Established: 1857
Features: Courtyard Atrio
Contact: Holy Family Church / (505) 351-4698

One of the things that make the Land of Enchantment so enchanting is its wealth of legends. Take, for instance, the story about a statue that walks around at night and the story about the people who cared for it.

It all started in Spain during the 13th century when Christians and Moors were battling each other. Following a fierce battle near the town of Atocha, the victorious Moors occupied the town and took many Christians prisoner. Except for small children, the captives were allowed neither food nor visitor. Fearing for the lives of their loved ones, the prisoner's family members besieged heaven with prayers and novenas asking God and the holy ones to provide some sort of help. On several different occasions, a small boy, wearing a wide-brimmed hat and carrying a gourd of water and a basket of bread, appeared at the prison. The Moors allowed the boy to enter so that he could feed the prisoners. Once all of the prisoners were fed, the boy quietly left with his basket and gourd still full.

Knowing neither whom the boy was nor where he came from, the good people of Atocha took it for granted that the Christ Child had come in answer to their prayers. From then on whenever anyone was in need of spiritual or physical sustenance they called upon El Santo Niño de Atocha, the Holy Child of Atocha.

Several centuries later in a little village known for its holy dirt and miraculous cures, Ramon Medina, who suffered from severe rheumatism, received a heavenly message telling him that if he prayed to the Santo Niño, he would be healed. Not only did he pray, Medina made a promesa (promise) that if he was cured, he would make a pilgrimage to the shrine of Santo Niño de Atocha in the Mexican town of Plateros. As might be expected,

Medina was cured, fulfilled his promise and, while in Plateros, was given a statue of Santo Niño to take home. Upon his return to New Mexico he built a capilla (small chapel) and dedicated it to the Holy Child.

The people of El Potrero were amazed by Medina's miraculous cure and, believing that the statue in Medina's capilla possessed the same power as that in Plateros, began visiting the chapel, placing candles, flowers, and small gifts at the statues feet. It was said the statue of Santo Niño de Atocha had a life of its own and, before long, rumors started to circulate throughout the small community.

Santo Nino Chapel Sign

"Did you see El Niño's shoes? They are muddy and worn out. Someone said they saw Him last night out near the Trujillo house. Young Juan was suffering with a fever but this morning he is well. Es una Milagro!"

Soon, tiny baby shoes were added to the ever-mounting collection of gifts laid before the statue. People from other villages began making pilgrimages to El Potrero in the hopes that their prayers would be answered. If they had no gift for the Holy Child, Sarita Medina, Ramon's wife, provided them with baby shoes, free of charge.

Throughout their lives, the Medinas remained devoted to their Santo Nino and His capilla. Every morning they opened the heavy wooden doors to let light and fresh air into the small chapel. They washed the windows, scrubbed the cracked linoleum floors, dusted the santos, and replaced Santo Nino's shoes whenever necessary. At night, they closed the doors, made sure the candles were safely snuffed out, and said goodnight to the Holy Child. Although Ramon and Sarita were not rich, they paid for the chapel's care from their own pockets, graciously accepting any donations left in a small metal box near the door.

Upon their deaths, only months apart in 1985, the Medinas were buried side-by-side in the chapel's small atrio (courtyard). Their shared grave sits beneath a lilac bush within a small wrought iron fence and is surrounded by plastic flowers left by friends and family to remember their old friends and tell visitors that the Medinas have gone to a better place. Yet to this day, some people swear they see shadows moving around the candle-lit capilla late at night.

In the trees around the chapel, a large black and white bird makes its presence known. Hovering around the graves, flitting from tree to tree, and hopping around on the ground, the Black-billed Magpie emits a distinctive, almost ear piercing, "yak-yak-yak. What is it trying to say? Were Ramon and Sarita seen scrubbing floors in the capilla or was Santo Niño spotted walking around the village again last night?

Are these legends true or just colorful examples of the stories heard around the Land of Enchantment?

Only the Magpie knows for sure.

Medina Graves in Atrio Outside Santo Niño Chapel

KIT CARSON PARK MEMORIAL CEMETERY

Location: 114 Kit Carson Road in Taos-north of the Plaza
Established: Around 1847
Features: Historic Gravesites
Contact: Kit Carson Park Office / (505) 758-2493

By the mid 1800s, life on the Northern Frontier was changing. Mexico fought for and won sovereignty of the fledgling territory, forts were built to protect remote settlements from marauding Indians, commerce along the Santa Fe Trail provided an increase in goods and materials, colonists and Pueblos had settled some their differences, and a burgeoning trade in beaver pelts brought a sudden influx of mountain men into the small village of Taos. One of those men was Christopher Houston Carson, better known as Kit Carson.

Born in Kentucky and raised in Missouri, Kit Carson left home when he was barely sixteen years old, hitched a ride on a wagon train, and set out for the untamed west. Arriving in Taos in 1826, he promptly hooked up with some trappers and began his legendary frontiersman life of hunting, trapping, and scouting.

Like many other frontiersmen, Carson spent a great deal of time among the Indians, possibly preferring an uncomplicated life to that of the more sophisticated "city" life in Taos. His first wife, an Arapahoe woman named Waa-nibe (Singing Wind), died during childbirth. A couple of years later, he married a Cheyenne girl who abandoned him so that she could travel with her tribe.

In 1842, Carson met up with a U.S. Army officer named John C. Fremont who, upon learning that Kit was fluent in several Indian languages, hired him as a guide. Fremont and Carson explored the Great Salt Lake, the South Pass of the Rocky Mountains, the Northwest Territory, and California. When the Mexican War broke out, Carson joined forces with the U.S. Army and fought at San Pasqual, San Miguel, and Los Angeles, emerging as a national hero and being commended for his actions by President James K. Polk. Journalists and writers quickly picked up on Carson's fame and, before long, tales of Kit Carson's adventures began appearing in the widely popular dime novels.

After sixteen years away from civilization, Carson decided it was time to settle down and, according to his memoirs, "enjoy the luxury of a meal consisting of bread, meat, sugar, and coffee." He and a friend bought some land, built an adobe house, and started a new life. In 1843, 14-year-old Maria Josepha Jaramillo became Carson's third wife and, in subsequent years, bore the aging frontiersman eight children.

Wanderlust, however, was in Kit Carson's blood. Whenever possible, he found his way back to the Rocky Mountains, chasing down renegades and outlaws, fighting predatory Indians, and herding sheep to market. He also served in the Civil War, guided wagon trains along the Santa Fe Trail, and turned his home into a stagecoach stop for weary travelers. In 1865, Carson was given a commission as a brigadier general and in 1866, he and his family moved to Colorado where he took command of Fort Garland.

Maria Josepha Jaramillo Carson died on May 23rd, 1868, ten days after giving birth to the couple's eight child, Josefita (Josephine). Christopher Houston Carson died one month later. Their bodies were returned to Taos where they were laid to rest in a small cemetery near their old home.

About the time Kit Carson first arrived in Taos, Padre Antonio José Martínez was taking over as the new parish priest. A New Mexico native, Martínez was born into a wealthy family in Abiquiu in 1793. Twelve years later, his family moved to Taos and built a hacienda. Although he never received a formal education, Antonio was a curious, intelligent young man who read everything he could get his hands on. When he was 19, he married and fathered a daughter. One year later, his wife died. His daughter died shortly thereafter. In 1817, Martinez moved to Durango where he studied for the priesthood. Five years later he was ordained as a Catholic priest. His first assignment was Tome (south of Albuquerque), followed by Abiquiu, Taos, back to Abiquiu, and then, finally, back to Taos.

Kit Carson Grave

Since Martínez believed all children, regardless of background, deserved a good education, he opened a school centered on religion, Latin grammar, arithmetic, language, and literature. The school was open to the children of ranchers, fur-traders, and natives alike. In the good padre's view, God was color-blind. The big problem, however, was there were more children than books. So Martínez bought a printing press and proceeded to print his own textbooks, catechisms, and missals. The printing press also came in handy when the priest published the first newspaper west of the Mississippi: El Crepusculo de Libertad (The Dawn of Liberty). Although the newspaper lasted only a short while, it served as the perfect platform for Martínez' libertarian views.

In an essay written by Steve Martínez we are told:

. . . much of Martinez' ideological position was centered on "Mexicano Patriotism and Republican civic values." In one example, Martínez expresses his political grievance by suggesting "Oh my mother land, for the sacred right of the governed people under the form of a Republican government? This is a gloomy time for New Mexico. Injustice, tyrannical laws and fees in New Mexico" are denying its people God's given rights . . . While Martínez courageously expressed grievances for his constituents, his patriotic duty to the Republic was reflected by his actions.

In 1853 when French-born Jean Baptiste Lamy became bishop of a newly formed diocese that included Taos, Martínez's radical views drew immediate attention. Not only was Martínez meddling in politics, he was known to have taken part in at least twelve minor rebellions, including the one which resulted in the execution of the New Mexico governor. But, horror of horrors, he was also a staunch supporter of that extremist group that called themselves "Los Hermanos." None of the priests Lamy brought from France would ever act in such a manner, but since there weren't enough European priests to go around, Lamy would have to find a way to restrain the actions of the native-born Martínez. Almost immediately, the two men became adversaries.

Over the next couple of years, letters were sent, visits were made, and admonitions were issued. The people of Taos loved their priest and did everything in their power to protect him. But, at the age of 59, Martínez decided enough was enough.

In January of 1856, Padre Martínez wrote to Bishop Lamy expressing concern that rheumatism, as well as several other ailments, might force him to retire. Lamy ignored the letter. Several months later, Martínez wrote a second letter suggesting a native-born priest that might replace him. This time, Lamy replied, accepted Martínez' resignation but notified the Taos clergyman that a Spanish priest, Don Damaso Taladrid, would be his replacement. Lamy had outmaneuvered Martínez.

Once Taladrid took over the Taos parish, he began bad-mouthing Martínez and everything the curate had accomplished. Martínez shot back and went as far as publishing an open letter in a Santa Fe newspaper delineating his altruism and Taladrid's malice. The war was on. Lamy traveled to Taos to try to bring about some sort of reconciliation between the two priests but no change took place so he went a second and then a third time. Still no change.

When it was discovered that Martínez had remodeled his residence to contain a private oratory in which he conducted private services for his congregation of natives and relatives, Lamy saw his chance. He issued a writ of suspension divesting Martínez' of all his priestly duties. Martínez defied the bishop and refused to terminate any of his ministrations.

In June of 1857, Bishop Jean Baptiste Lamy set into motion formal proceedings to excommunicate Padre Antonio José Martínez. Parishioners tried to fight the excommunication but Father Martínez slipped into spiritual exile, continuing with his illicit church until his death on July 27, 1867. He was buried beneath the cottonwood trees in the same cemetery as Kit Carson.

Over the years, Martínez's grave became the target of vandals. In 2004, the desecrated headstone was removed and replaced with a new one.

Mabel Dodge Luhan never knew Carson or Martínez, but she had a lot in common with both of them. She, like they, loved Taos, she preferred doing things her way, and she never took the easy way out.

Born on February 26, 1879, Mabel Ganson and her family were considered the "elite" of Victorian Buffalo, New York. As a child,

Padre Martinez Grave

Mabel had a nursemaid; as a young girl, she attended a private girls' school; and, at the proper time, she was given a coming-out party. She traveled extensively with her family throughout Europe, learned several languages, and developed a penchant for fine art and literature. But she wasn't happy.

Mabel married for the first time at the age of 21. It was, in her words, "a passive, truly feminine experience." Mabel's only child, John, was born in 1902. In her memoirs she recalled, "It seemed I didn't want a baby after all." What Mabel Ganson Evans wanted was to "find" herself.

Following the unfortunate death of her husband who was killed in a hunting accident, Mabel Evans, her infant son, and two trained nurses, set out for Europe. On the voyage over she met Edwin Dodge, a young Bostonian architect, who pursued Mabel until she finally agreed to marry him. But this marriage wasn't any happier than the first. In fact, Mabel felt "quite desperate and with no heart at all." To compensate, she started buying things. "My thoughts were of a life made up of beautiful things, of art, of color, of noble forms, and of ideas and perceptions about them that had been waiting, asleep within me, and that now allured me by their untried, uncreated images." She decided to become a "Renaissance Lady."

Edwin and Mabel purchased a fifteenth century Florentine villa to house all the "objet d'art" that Mabel had purchased. Sculptors, artists, writers, actors, and various members of the international set supped at Mabel's table. It seemed Mabel had everything she ever wanted. But before long, the veneer of Florence's glitz and glamour wore thin. Mabel wanted something more. So she became "la otra mujer"—the other woman.

It's not known exactly how many men Mabel Dodge "entertained" in her Florence mansion. It is known, however, that husband Edwin didn't take kindly to his wife's activities. After watching a succession of suitors come and go, he offered Mabel a choice: either change her ways and return with him to America or consider their marriage ended.

Although Mabel returned to America, she didn't change her ways. This time she turned to her chauffeur for comfort. But the chauffeur, stricken with anxiety couldn't, or wouldn't, respond. Mabel's subsequent attempted suicide showed her that she had been "playing house" for too long and that it was time she took her place in respectable society as a mentor.

Mabel, Edwin, and son John moved to New York where Mabel hoped to begin her new life. Unfortunately, that new life was put on hold when she suffered a nervous breakdown. Following her recuperation in Europe, Mabel returned to New York, met Alfred Stieglitz, "the center of avant-garde art in America," and began work on an art show that would "dynamite New York." The show was a great success. A short while after the show, Mabel met and fell in love with John Reed, a man who enjoyed literature, poetry, politics, and, of course himself. The affair, naturally, was doomed. By 1915, Mabel had a new lover, Maurice Sterne. On June 8th of 1916, Mabel received word that she and Edwin were divorced and on August 23, 1917, she and Maurice were married.

In the early 1900s, writers and artists, including Ernest Blumenshein, Alfred Stieglitz, Georgia O'Keefe, and Mary Austin, discovered that New Mexico was "The" place to be. Consequently, when Mabel witnessed Maurice looking at another woman just a few short months after their marriage, she sent her new husband off to the Southwest to "check things out." Sterne set up a studio in Santa Fe but, in one of his many letters to Mabel, he wrote that "he could not work without her." After a grueling cross-country train trip, Mabel arrived in Santa Fe only to find that the city was "too tame and conventional" for her tastes so she decided to join friends already involved in the Taos Society of Artists.

Taos was everything Mabel hoped it would be. The sky seemed bluer, the colors deeper, the air fresher, and the people friendlier. One of Mabel's friends, Andrew Dansburg, once said "Taos has the quality of a place in which...to find God." Mabel may not have found God but, in Taos, she found a place where she could satisfy her artistic and spiritual hungers and she found Tony Luhan, a Pueblo Indian.

Tony introduced Mabel to Pueblo customs, the mystical beauty and inspirational character of Taos, and the meaning of love. He also pledged to give her "a new life—a new world." In December of 1922, Mabel divorced Maurice and in April of 1923, she married Tony.

The house that Mabel and Tony renovated reflected Mabel's new world. It was bright. It was open. It was large. Mabel had windows put in everywhere so that the Taos light that everyone so admired would fill the rooms. However, when D.H. Lawrence visited Mabel, he was so appalled by the lack of privacy in the bathroom that he painted over the windows. So much for Taos light.

Needless to say, Tony and Mabel did not live happily ever after. There were rumors about Tony and other women, Mabel and other men, and Mabel and other women. Some were true; some were not. Regardless, Mabel and Tony stayed together for more than forty years, opening their home to artists, actors, and scientists, traveling around the world, and sharing their unique life-style with anyone who was interested.

Mabel Ganson Evans Dodge Sterne Luhan died on August 13,

1962 and, following a simple ceremony, was buried in the Kit Carson Cemetery (a man she hated because he killed Indians). Tony died a year later and was buried in the Taos Pueblo graveyard.

Mabel Dodge Luhan Grave

The Kit Carson Memorial Cemetery is located in a quiet corner of Kit Carson Park just north of the Taos Plaza. There are only thirty or forty graves there, mostly Carson family members, Padre Martínez and Mabel Luhan. Somehow, it seems appropriate that Mabel, a renowned eccentric and promoter of Indian culture, Padre Martínez, known for his crusades, political activism, and distrust of outside church authority, and Kit, a frontiersman, trapper, scout, soldier, and family man are all buried near each other. Although only a few of the movers and shakers that lived in this small village, they epitomize the strength and determination of those who helped create the mystique that sets Taos apart from the rest of the state, and very possibly, the rest of the world.

D.H. LAWRENCE MONUMENT AT KIOWA RANCH

Location: San Cristobal, New Mexico halfway between Taos and Questa on Highway 522
Established: 1934
Features: Shrine and Cabins
Contact: University of New Mexico / (505) 277-6248

English born David Herbert Lawrence, world-renowned poet, essayist, and novelist, once wrote: *I think New Mexico was the greatest experience from the outside world that I have ever had. It certainly changed me forever.*

And well it may have because up until the time he and his wife Frieda first visited New Mexico, Lawrence's life was in constant turmoil.

Following an unsatisfying teaching career and already suffering from tuberculosis, Lawrence eloped to Germany with Frieda von Richtofen Weekley, daughter of the German Baron von Richtofen and then-wife of Ernest Weekley, one of Lawrence's professors. Already receiving criticism for his evocative prose and poetry, D.H. and Frieda's controversial relationship drew attention at the outbreak of World War I when people accused them of being German sympathizers and spies. The couple moved from country to country, looking in vain for peace and privacy. In 1922, Mabel Dodge invited them to Taos.

Lawrence fell in love with New Mexico. Even though he only stayed a short while, he decided to create a utopian society in Taos and asked several friends to move there with him and Frieda. Only one, English painter Dorothy Brett, agreed and in 1924, the three friends moved to a 160-acre ranch given to Frieda by her friend Mabel Dodge.

Kiowa Ranch, so named because of an Indian trail that ran through the property, was located on Lobo Mountain about 20 miles northeast of Taos. John Craig homesteaded the land in the late 1880s but sold it to Mary and William McClure in 1893. The McClures raised Angora goats on the property until selling it to Mabel Dodge in 1920. But Mabel's life was in almost as much turmoil as D.H.'s so when she discovered that the Lawrences were thinking about permanently moving to Taos, she gave them the Kiowa Ranch.

D.H. spent the summer of 1924 repairing and remodeling the two cabins on the ranch. During that time, he completed a short novel "St. Mawr" that extols the aesthetic beauty and near-spiritual character of Kiowa Ranch. The summer of 1925, however, was the last time Lawrence would visit his beloved Kiowa Ranch.

D.H. Lawrence Memorial Chapel

Seeking treatment for his long-neglected illness, D.H. and Frieda moved to Vence, France where D.H. entered a sanatorium. He died in March of 1930 and was buried in the local cemetery. Frieda moved back to Kiowa Ranch and five years later she and her new husband, Angelo Ravagli, had Lawrence's remains exhumed, cremated, and shipped to the United States. When Lawrence's ashes arrived in Taos, a disagreement broke out between Frieda, Mabel, and Dorothy Brett. Although Frieda intended to have the ashes interred in a small memorial she built in honor of her late husband, Mabel and Dorothy, both openly in love with D.H., insisted his ashes be scattered across the ranch. Thinking quickly, Frieda poured Lawrence's ashes into wet cement and had Ravagli build an altar inside the memorial. Frieda died in 1956. Her grave is outside the entrance to the memorial.

During his lifetime, the words of D.H. Lawrence were denigrated, censored, and banned. Some people thought his work was too provocative; too sensational; too sexual. He was rebellious, radical, and profoundly polemic. Yet he always seemed to get to the heart of a problem whether it involved the industrialized society, nature, sex, or just plain living. He put down on paper what others feared to think. Maybe his greatest crime was seeing the world as it really was.

Over the years, Kiowa Ranch and the Lawrence Memorial fell into disrepair. A sign directing visitors to the memorial was disfigured with graffiti, windows were broken in the cabins, and the shrine with Lawrence's ashes was almost in ruins. The University of New Mexico (to whom the property now belonged) deliberated whether or not to make repairs. There was the argument that the ranch was too far from the university's campus, that it would cost too much to make repairs, and that it just wasn't a priority. Obviously, other people didn't agree. Repairs were made and Frieda's dying wish to insure that the ranch was maintained and opened to the public was fulfilled.

Today, visitors drive a three or four mile dirt road to reach Kiowa Ranch. Once there, they peer through rain-spattered windows trying to get a glimpse at D.H.'s old typewriter or denim jacket, and they walk a zigzag path to a white shrine set on a hilltop above the cabins. Inside the shrine, there is evidence of two legged and four-legged visitors. The four-leggeds leave their distinctive little messages—this is the country, after all—and some of the two-leggeds leave offerings and sign a guest book. One visitor wrote:

What he had seen and felt and known, he gave in his writing to his fellow men—the splendor of living, the hope of more and more life—he had given them a heroic and immeasurable gift.

D.H. Lawrence Shrine

The Phoenix that died in flames and was reborn from its own ashes was D.H. Lawrence's personal idol. This mythical bird of old used its magical powers to create excitement and dispel enchantments. Lawrence used words to do the same thing. He was a writer of words, a seeker of truth and justice, and he brought light into a world of darkness. Statues of the Phoenix rise above the outside roof and the inside altar. It's good to know that he found peace and tranquility in the hills high above Taos.

VIETNAM VETERANS MEMORIAL STATE PARK

Location: In Angel Fire on Highway 64 East
Established: Dedicated in 1971
Features: Visitors Center and Monument
Contact: Park Office / (505) 377-6900

As you climb the twisted, forested road between Taos and Angel Fire, you might feel as if you are ascending into heaven. Somehow, the air seems crisper, the skies bluer, the clouds fluffier, and the trees taller than any you have ever seen. Even the birds sound happier. After about twenty miles, the road drops down into a narrow valley and, suddenly, it comes into view—the Vietnam Veterans Memorial.

The vast gull-like structure rises above the brow of a knoll to a height of nearly 50 feet and has graceful, inward curving walls sweeping down to each side of twin center pinnacles. The west wall is slightly higher and longer and is a quarter-circle arc with a 99-foot radius. Both walls flow majestically down from their commanding height so that the tip of each disappears as it is buried in the ground. A third inward curving wall completes the structure.

This is how Dr. Victor Westphall, founder of the memorial, described the structure he created, not to glorify war, but to honor the men and women who served in Vietnam and acknowledge the selfless sacrifices they made for their country. For those who see it for the first time, the snow-white building evokes the image of a white dove hovering over the little village below.

Plans to construct the memorial began five days after Dr.Westphall and his wife learned that their son, Marine Lt. Victor David Westphall III, died in an enemy ambush in Quang Tri Province on May 22, 1968. With little more than their son's military life insurance, Dr. Westphall, his wife Jeanne, and their younger son Douglas set out to create the United State's first memorial to honor all the men and women lost or wounded (physically was well as mentally) during Vietnam War. The Westphalls called their project the Vietnam Veterans Peace and Brotherhood Chapel.

Vietnam Veterans Memorial

Funds were scarce and the building season short, so we did not complete the structure until the spring of 1971. Even so, the design was so stunning, and its purpose so compelling, that it was already drawing national attention. We held the formal dedication on May 22, 1971, the third anniversary of David's death.

As news of this remarkable memorial spread, visitors began pouring in from all parts of the country. Veterans on motorcycles, hikers with pack packs, tourists with cameras, school children, artists, and those seeking closure braved the road and unpredictable weather to find solitude and answers at this place on the hill. Georgia O'Keefe visited the Memorial shortly after it opened. Her biographer, Jeffrey Hogrefe, described the artist's visit:

It was cold outside. The field around the memorial was muddy, and boards had been placed over the mud. Some of the boards had slipped and turned sideways. (Juan) Hamilton (Georgia's companion and caretaker) lifted O'Keefe from the car and carried her across the boards to where she could reach the wall. The woman who had vehemently opposed the First World War looked at it sadly. She could not see, but she could feel. Hamilton hoisted her closer to the wall. She placed her hand on one of the names and fingered the indentation in the stone where a name had been etched. She began to cry.

Within the first ten years, so many people visited the memorial that it soon became apparent that additional facilities were necessary. However, plans to add a visitor's center ran into numerous roadblocks—lack of support, broken promises, increasing costs to maintain the original memorial, and decreasing funds. For the most part, funds for maintenance and construction came from the Westphall's personal resources and sporadic private contributions. The future of the memorial looked dim.

In 1981, the DAV Vietnam Veterans National Memorial was organized as an entity separate from the Disabled American Veterans Organization. While the new entity was responsible only to itself, the parent organization came to be most generous in all facets of its support. The new venture would grow into a $2 million creation commanding the attention of our Nation. At an organization meeting at Angel Fire on September 6, 1982, title to the Vietnam Veterans Peace and Brotherhood Chapel was conveyed to the DAV Memorial.

Construction of a new visitor's center began in 1984 and was completed in 1986. The following year, President Ronald Reagan signed a proclamation recognizing the DAV Vietnam Veterans Memorial as a memorial of national significance. In 1998, the DAV returned title of the Memorial Complex to the David Westphall Veterans Foundation but rising costs and declining funds once again threatened the future of the memorial. At the end of the State of New Mexico's 2004 legislative session, the House of Representatives unanimously approved Senate Joint Resolution 11, approving state acquisition of the National Vietnam

Veterans Memorial. At 11am on November 11th of 2005, the memorial became New Mexico's 33rd state park. It is the only state park in the U.S. dedicated solely as a vietnam veterans memorial.

Today, as in years past, the mood at the memorial is reflective but not morose. Walking around the grounds, most people speak in whispers, if at all. Inside the chapel, they pray for the POWs, MIAs, and all those wounded and killed in this and all other wars. And they pray for peace.

"Doc" Westphall, as he came to be known, passed away on July 22, 2003. Surrounded by trees, shrubs, native grasses, and wild flowers, his final resting place is atop the hill just behind the chapel. It would be nice to think that Doc's spirit joined those of the hundreds of thousands lost or wounded in the Vietnam War and that, together, they soar on the snow-white wings of a white dove high above the monument dedicated to their sacrifices.

3

SANTA FE AREA

THE MORMON BATTALION MONUMENT

Location: Budaghers I-25 Exit 257 between Santa Fe and Albuquerque
Established: 1940—Relocated in 1997
Features: 25-Foot Stone Obelisk
Contact: Mormon Battalion Association, Sandy Utah, 801-256-0329 or www.mormonbattalion.com

They were farmers and fathers and husbands yet they undertook a task that other men would not. Armed with little more than picks, shovels, and prayer books, a group of 543 men, 34 women, and 51 children left Council Bluffs, Iowa on July 20th, 1846, on an historic 1850 mile march to San Diego, California. Known as the Mormon Battalion, these hardy souls had volunteered to fight for the United States in the Mexican War in exchange for assistance during their relocation to the Great Salt Lake Valley in Utah. Their lives back home were dismal so they had nothing to lose and everything to gain. Little did they know the pain and suffering they would experience on what would later be deemed the nation's longest military march.

The Mormon Church, originally known as "The Church of Christ," was founded by Joseph Smith in Fayette, New York on April 6, 1830. In 1831, the church was moved to Kirtland, Ohio and in 1832, some Mormons started to settle in Missouri. Because of their unconventional beliefs and practices, the new settlers were ill received. War broke out between believer and non-believer, eventually leading to the death of Smith in1844 and the selection of a new leader, Brigham Young, who spearheaded the search for a place where the Mormons could practice their faith in peace. Such a place was in Utah but, to get there, the pilgrims would have to cross Indian country. By late 1845, they were only as far as Iowa.

In January of 1846, Young wrote to President James K. Polk, offering to build block houses and stockade forts along the westward trails in exchange for assistance for his migrant followers. His reply came in June when, following the United States' declaration of war against Mexico, Captain James Allen rode into the Mormon camp asking for volunteers to fight the Mexicans in California. Young's answer was quick. "You shall have your men and if we have not enough men, we will furnish you women." Allen led the group of rag-tag soldiers from Iowa into Kansas where, following his sudden death, Lieutenant A.J. Smith took command.

Most of the Mormons considered Smith a cruel and demanding leader. Complaints about inadequate food, improper medical care, and extraordinarily long forced marches were frequent. Luckily, once the battalion reached Santa Fe, Smith was relieved of duty and replaced by Lieutenant Colonel Phillip St. George Cook who took the Mormon soldiers to their destination. Upon arriving in San Diego, Cook paid tribute to these men in part as follows:

History may be searched in vain for an equal march of infantry. Half of it has been through wilderness, where nothing but snakes and wild beasts were found, or deserts, where for want of water, there is not a living creature. There, with almost hopeless labor, we have dug wells, which the future traveler will enjoy. Without a guide who had traversed them, we have ventured into trackless tableland, where water was not found for several marches. With crowbar and pick and axe in hand we have worked our way over mountains which seemed to defy aught save the wild goat and hewed a pass through a chasm of living rock more narrow than our wagons. To bring these first wagons to the Pacific, we have preserved the strength of our mules by herding them over large tracts, which we have laboriously guarded without loss. Thus, marching half naked and half fed and living upon wild animals, we have discovered and made a road of great value to our country.

Following their discharge from military service, the soldiers and families of the Mormon Battalion went on to join other pilgrims already in the Valley of the Great Salt Lake. Over the next 22 years, nearly 70,000 pioneers followed.

In recent years, numerous monuments and trail markers have been placed along the route followed by the Mormon Battalion volunteers. The monument at Budaghers is of particular interest because of its location. While others are near or in cities, this monument is situated in an isolated field approximately 25 miles south of the hoopla and frenzy of downtown Santa Fe. Before the building of Interstate-25, the monument was positioned several miles north of its present location but "progress" forced it to be moved "out of the way." For a while, the future of the original monument was uncertain until a committee of New Mexican residents resituated it near the hamlet of Budaghers.

Mormon Battalion Monument

The present location is appropriate because it conveys the very essence of the Mormon Battalion. Battered by recurring storms and strong winds, cut off from the rest of the world by the surrounding wilderness, frequented by snakes, scorpions, and coyotes, the stone obelisk exemplifies the strength and courage of the men and women of the Mormon Battalion who, though they never fought in a war, opened routes of transportation into California, helped to settle the entire northern Mexico region (including New Mexico), played an important role in the discovery of gold at Sutter's Mill, and left friends and family behind to walk half way across the continent to help defend the United States.

They more than earned their place in history.

SANTUARIO DE GUADALUPE

Location: 100 S. Guadalupe Street, Historic Guadalupe District—Santa Fe
Established: Mid to Late 1700s, possibly earlier
Features: Our Lady of Guadalupe Reredo and Re-interred Burials
Contact: 505-988-2027

Although there are undocumented claims that this landmark church dates back to1640, the consensus of opinion is that it was built sometime between 1776 and 1795. Originally known as Nuestra Señora de Guadalupe, it was the last place travelers stopped before embarking on their journeys along El Camino Real, the trail linking Mexico City with its northern provinces. It was here they said prayers for a safe passage, took water from the Aqua Fria well, and kissed their loved ones good-bye.

According to church records, *the original church was a flat-roofed, dirt-floored wood and adobe structure surrounded by an adobe wall and surmounted by a three-tiered bell tower that housed locally produced copper bells.* During most of its early years, this church, situated on a slight rise above the Santa Fe River, was without an assigned priest. Services were infrequent, few if any records were maintained, and upkeep was neglected.

When the railroad arrived in New Mexico in 1880, it brought an influx of new people with new ideas. One of those people was Father James H. DeFouri who was assigned by Archbishop Jean-Baptiste Lamy to oversee renovation of the deteriorated church and turn it into a place where new arrivals could worship.

Like Lamy, DeFouri preferred stately, gothic architecture to antiquated, rustic designs. The church's original bell tower was demolished and replaced by a Colonial-style wooden spire, the adobe wall around the church was torn down and replaced by a picket fence, windows were cut through the massive 33-inch-thick walls, the roof was changed from flat to pitched, rows of wooden pews were added, and the name was anglicized to the Chapel of Guadalupe. About the only thing that wasn't changed was the reredo (altar screen) that graced the otherwise unadorned altar.

Painted in 1783 by artist José de Alzibar, one of Mexico's most renowned painters, the screen consisted of six individual oil paintings, four representing scenes relating the story of the Virgin of Guadalupe, one depicting the three elements of the Trinity as humans rather than spiritual figures, and the central painting portraying the Virgin as she appeared on the cloak of the shepherd. The six paintings were transported by mule from Mexico City to Santa Fe where the 14-foot high, 10-foot wide screen was assembled and placed on the altar.

Reredo at Santuario de Guadalupe

A devastating fire gutted the church in 1922, taking most of the roof but leaving the reredo unharmed. In rebuilding the church, a new design emerged—that of California Mission Style. Stained glass windows were added and, for a while, it appeared the church had been given a new lease on life.

By the 1960s, the congregation had outgrown the old church and replaced it with a bigger, better, more modern structure. The old church was closed and, once again, fell into disrepair, so much so that there was talk of razing it and turning the land into a parking lot. In 1975, the Guadalupe Historic Foundation was founded, most of the previous renovations were reversed, and the Santuario was returned to its original appearance and dignity.

In the late 1980s, a Santa Fe contracting firm began work to stabilize the church's bell tower. During the process of digging a new foundation, upwards of 80 graves were discovered beneath the floor. A team of archaeologists was called in to examine the graves and it was determined that the remains dated back to the mid-1800s when church floor burials, although no longer permitted, were still being practiced. After consulting with the Archbishop, the remains were removed to a safe place. Following completion of the construction work, a religious service was performed, and the remains were re-interred in a special crypt under the floor where they were discovered.

Throughout the years, the church that stands in the heart of Santa Fe's Historic Guadalupe District has survived many changes. At one time, it looked as if it belonged in Colonial New England; at another, it took on the California Mission-Style. It was modernized; it was burned; it was abandoned. No longer a consecrated church, the Santuario lives life anew as a museum and performing arts center. However, even though people come here to listen to concerts or browse through old books and documents, they leave knowing they have been in a place of goodness and grace and they remember the Santa Fe that once was.

Graves Along Foundation at Santuario de Guadalupe

THE CATHEDRAL BASILICA OF SAINT FRANCIS OF ASSISI

Location: 131 Cathedral Place—Santa Fe
Established: First church 1610, Rebuilt 1714, Present church 1886
Features: Wall burials and Floor Crypt
Contact: Cathedral Office / (505) 982-5619 or
www.archdiocesesantafe.org

Cathedral Basilica of St. Francis of Assisi

When Santa Fe, the city of Holy Faith, was founded in 1610, two churches were built: one for the natives, one for the Spaniards. During the Pueblo Revolt of 1680, the Indian church, the church of San Miguel, was severely damaged and the Spanish church, the church of St. Francis of Assisi, was destroyed. When the Spanish colonists returned to Santa Fe in 1693, they repaired the church of San Miguel and completely rebuilt the church of St. Francis in 1714.

In 1759, after completing construction of a military church known as the Castrense, Governor Marin del Valle made two trips to recover the remains of two Franciscan friars. According to L. Branford Prince's 1915 book, *Spanish Mission Churches of New Mexico*:

Both were carried back to the capital, and on August 31, 1759, *were buried in a large coffin which was placed in the wall of the Gospel side of the parish church, where it still remains.*

On the coffins are two inscriptions in Spanish, of which we give English translations, as follows:

"Here rest the bones of the venerable P. Fray Geronimo de la Llama, an apostolic man of the order of St. Francis. These bones were unearthed from the ruins of the old Mission of Quarac in the Province of Las Salinas on April 1st, 1759.

and

"Here rest the bones of the venerable Fray Ascencion Zarate, an apostolic man of the order of St. Francis. These bones were exhumed from the ruins of the church of San Lorenzo of Picuris, on May 8th, 1759; *and the remains of the two venerable missionaries were transferred to this Parish of Santa Fe and were buried on August* 31st, *of the same year,* 1759."

The church referred to as the "parish church" was the Spanish church known as the church of St. Francis of Assisi. The coffin was placed in a wall near the Chapel of Our Lady La Conquistadora, a chapel built to shelter the statue of the Virgin Mary brought to Santa Fe in 1626, carried to safety during the years of the Pueblo Revolt, and returned to the city when the settlers returned in 1693.

When Jean-Baptiste Lamy, a Frenchman, became bishop and took charge of the church in 1851, he embarked on a complete renovation of the 1714 church that would, according to artist and writer Marie Romero Cash, "relieve Santa Fe of its crude pueblo look and give it badly needed sophistication."

Statue of Archbishop Lamy in Front of the Cathedral

Lamy, however, knew that he couldn't deprive his congregation of their church therefore, instead of tearing down the old church and building anew, he instructed workers to erect a new structure over and around the old. Once complete, several years after Lamy's retirement, a large Romanesque-style stone structure stood where a humble adobe had once been. All that remained of the 1714 church was the Chapel of Our Lady, La Conquistadora, and the coffin of the Franciscan friars in an adjacent wall.

When Lamy died in 1888, a new tradition began. Although church floor burials had previously been prohibited by Spanish rule, by 1846 New Mexico was no longer under Spanish rule and the practice, at least in the case of Archbishops, was revived.

Commemorative Plaque Inside Cathedral

Beginning with Lamy and continuing throughout the years, many of Santa Fe's Archbishops have been buried in a crypt below the Cathedral floor. A plaque near the sanctuary lists the names of those buried there:

John Baptist Lamy
Peter Bourgarde
Albert Thomas Daeger, OFM
Rudolph Aloysius Gerken
Edwin Vincent Byrne
James Peter Davis

The Cathedral Basilica of Saint Francis of Assisi is a grand structure, adorned with imported stained glass windows, lit by frosted glass chandeliers, and sustained by massive Corinthian columns that lead to the vaulted ceiling above. Yet, for all of its grandness, this is a church of people who have forgotten neither their simple beginnings nor their reason for existence.

CROSS OF THE MARTYRS WALKWAY

Location: Paseo de Peralta at Otero Street—Santa Fe
Established: 1977
Features: Memorial Cross and Informative Plaques
Contact: Santa Fe Visitors Bureau—800-777-2489, 505-955-6200, or www.santafe.org

The arrival of 16th century Spanish explorers and colonists in the northern province of New Mexico was a double-edged sword for the Pueblo people who lived there. On the one hand, the colonists offered seeds for a variety of new food crops, tools to plant and harvest the crops, and the promise of a better life. In this arid region, rain was infrequent, water was scarce, and crops were hard to grow. The Indians knew any help they could get might mean their salvation. On the other hand, the newcomers demanded tribute, acceptance of an alien religion, and complete compliance to their way of life. What the Spaniards demanded greatly overshadowed what they offered but by the time the Pueblo people realized what was happening, it was too late. Their food supplies had significantly dwindled, other Indian tribes were pillaging Pueblo lands in search of Spanish wealth, and the native people had grown dependent on their oppressors.

The missionaries who accompanied the Spanish colonists were Franciscan friars. As part of the colonization process, one or more friars were assigned to each of the Pueblo villages for the sole purpose of spreading the Catholic faith and building churches, most of which was accomplished through intimidation and abuse. If the natives wanted food, they had to reject their old beliefs, be baptized, and help build the churches. Resisters that continued to participate in banned activities like singing, dancing, or taking part in traditional ceremonies were punished, starved, or sold into slavery.

After almost 100 years of subservience, deprivation, warfare, and persecution, the time came when the Pueblo people said "enough." And so, in 1680, they planned and executed a revolt aimed at running the Spaniards, including the friars, off their land. During the revolt, soldiers and colonists were killed or driven off, farms were ransacked, crops were destroyed, churches were burned, and 21 Franciscan friars were martyred. Twelve years later, however, the spaniards returned. And, in 1712, Governor Marquez de La Puñuela signed a proclamation establishing the 14th of September as a festival day to commemorate that return.

Known as La Fiesta de Santa Fe, the festival was formed to pay tribute to General Don Diego de Vargas Zapata Luján Ponce de León and Marques de la Nava de Brazinas, the conquistadors who triumphantly reentered and reoccupied Santa Fe. The 1712 Proclamation stipulated that "the day be celebrated with Vespers, Mass, Sermon, and Procession through the Main Plaza." Today, almost 300 years later, the Fiesta has achieved status as the *Oldest Continual Community Celebration in the United States* and features events such as an Arts & Crafts Show, selection of a Queen and her court, Mariachi concerts, several parades, the *Entrada* (re-enactment of de Vargas's peaceful entry into the city of Holy Faith), and the closing ceremonies that include a Mass of Thanksgiving and a candlelight procession to the top of a hill in Fort Marcy Park. Along the walkway leading up the hill, bonfires burn brightly as pilgrims make their way to the top where a 20-foot white cross stands. At the base of the cross is a plaque listing the names of the 21 Franciscan friars martyred in the Revolt and the places where they died.

The Cross of the Martyrs was constructed by Santa Fe High Vo-Tech students and erected by the American Revolution Bicentennial Commission and the Santa Fe Fiesta Council in 1977. The Walkway, constructed in 1986, was a 375th Anniversary of Santa Fe gift to visitors and features 20 informative plaques that summarize the early history of the city and the events that led to the death of the friars.

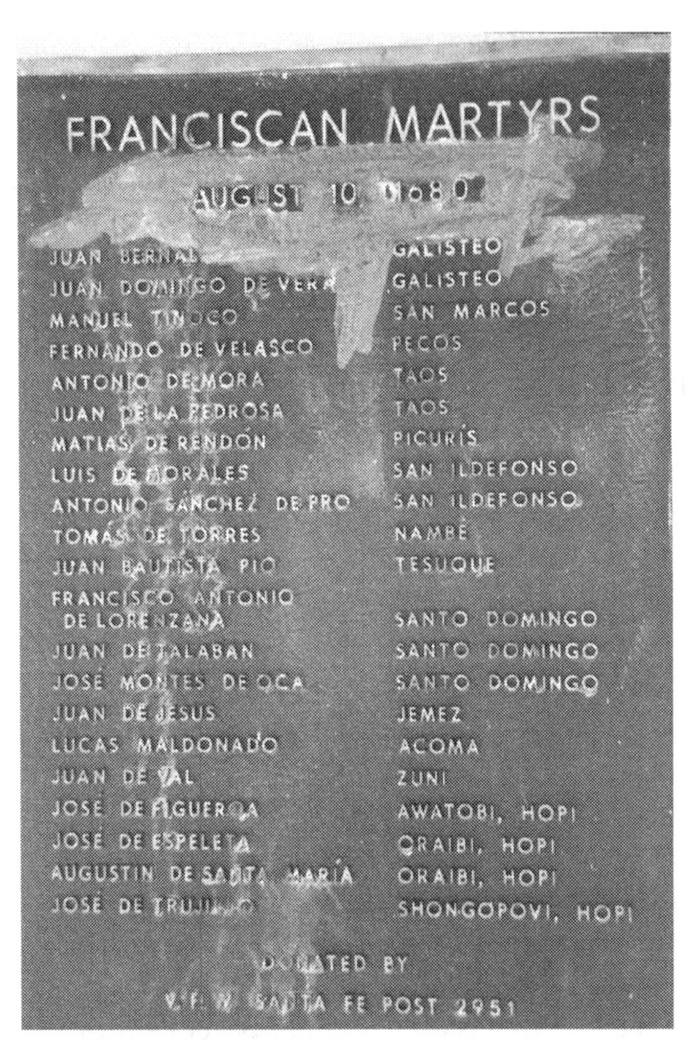

List of Franciscan Friars Killed During Pueblo Revolt

Juan Bernal	Galisteo
Juan Domingo de Vera	Galisteo
Manuel Tinoco	San Marcos
Fernando de Velasco	Pecos
Antonio de Mora	Taos
Juan de la Pedrosa	Taos
Matias de Rendon	Picuris

Luis de Morales	San Ildefonso
Antonio Sanchez de Pro	San Ildefonso
Tomas de Torres	Nambe
Juan Bautista Pio	Tesuque
Francisco Antonio de Lorenzana	Santo Domingo
Juan de Talaban	Santo Domingo
Jose Montes de Oca	Santo Domingo
Juan de Jesus	Jemez
Lucas Maldonado	Acoma
Juan de Val	Zuni
Jose de Figueroa	Awatobi, Hopi
Jose de Espeleta	Oraibi, Hopi
Augustin de Santa Maria	Oraibi, Hopi
Jose de Trujillo	Shongopovi, Hopi

The Pueblo people and the Spanish colonists eventually came to terms and formed an equitable, though tenuous, alliance. Both sides committed great errors but they learned from their mistakes and went forward. Yet, whenever someone, whether Indian or European, climbs to the top of the hill and looks at the cross, they remember the lives lost in colonizing New Mexico. And, they wonder . . .

Cross of the Martyrs

ZOZOBRA

Location: Fort Marcy Park—Santa Fe
Established: 1926
Features: Effigy Burning
Contact: Santa Fe Kiwanis Club (505) 660-1965
or zozobra.com

Another event that takes place during the Fiesta in Santa Fe is the burning of an ugly 50-foot tall ogre known as Zozobra or Old Man Gloom. The gist of this spectacular demonstration is that when Old Man Gloom is burned, everyone's cares and worries go up in flames with him.

The burning of effigies probably dates back to the time of the Inquisition when burning at the stake was the preferred form of punishment for immoral behavior. In those days, witches, warlocks, harlots, and other sinners often found themselves tied to a pole with piles of wood laid at their feet and hecklers casting jeers, rocks, and rotten eggs toward them. The fire was a form of purification that, obviously, cured the reprobates of their transgressions. However, when it was decided that burning was too strong a punishment, effigies were used as a means of changing the evildoers or bringing an end to their powers.

In later years, effigies became the focus of many festivals and rituals including the Hindu burning of Ravana the enemy of Lord Rama, Wicker Man a springtime pagan ritual believed to cleanse the earth, the Yaqui Indian symbolic burning of Judas during Easter week, Guy Fawkes Day when Brits burn their country's most notorious traitor, the hedonistic Burning Man Festival in Nevada, and, of course, football and political rallies where adversaries meet their fate at the hands of their opponents.

In 1924, Santa Fe artist Will Shuster decided to create a little excitement for his guests at a private party he held in his home during that year's Fiesta. People had been saying that Fiesta had become "dull and commercialized" so Shuster created a Yaqui-like effigy figure, filled it with firecrackers, and set it off during the party. One of Shuster's friends, newspaper editor E. Dale Johnson, nicknamed the creature "Zozobra" which in Spanish means "Gloomy One." Everyone had so much fun at the party that Shuster recreated his "grouch" the following year.

Zozobra

In 1926, Shuster, Johnson, and several Santa Fe artists and writers got together to produce a "revolutionary protest fiesta" called Pasatiempo (the Pastime) that ridiculed the then-tedious traditional Fiesta. Part of their mock celebration included Shuster's "Zozobra"—a burlap covered effigy stuffed with excelsior and soaked in a copper sulphate solution that created green flames when it was ignited. The September 2nd edition of *The Santa Fe New Mexican* described what took place:

Following vespers at the Cathedral, a long procession headed by the Conquistadores Band marched to the vacant space back of the city hall, where Zozobra, a hideous effigy figure 20 feet high, produced by the magic wand of Will Shuster, stood in ghastly silence illuminated by weird green fires. While the band played a funeral march, a group of Kiwanians in black robes and hoods stole around the figure, with four others seated before the green fire.

When City Attorney Jack Kennedy . . . solemnly uttered the death sentence of Zozobra . . . and fired several revolver shots at the monster, the green fires changed to red, the surrounding ring of bonfires was ignited, red fire blazed at the foot of the figure and shortly a match was applied to its bases and leaped into a column of many colored flames.

As it burned, the encircling fires blazed brighter, there was a staccato of exploding fireworks from the figure and round about, and throwing off their black robes, the spectators emerged in gala costume, joining an invading army of bright-hued harlequins with torches in a dance around the fires as the band struck up "La Cucaracha."

Zozobra was such a success with the crowd that it became a tradition in each of the following Fiesta celebrations.

Throughout the years, several changes were made to Old Man Gloom. One year, he took on the grimace of Hirohitimus—a bizarre combination of Emperor Hirohito, Adolph Hitler, and Italy's dictator Benito Mussolini. Another year, he looked like a creature from an alien planet. Except for the years of World War II when he shrank to eight feet, Zozobra grew taller every year. Stuffed with shredded documents like old police reports, paid off mortgage papers, and divorce decrees, he now reaches a whopping 50 feet. Always dressed in white muslin, his belt and buttons sometime change colors as does his fuzzy hair. But, regardless of any changes made, nothing affects his ghoulish demeanor. In fact, according to A.W. Denninger, Kiwanis Club Member and volunteer Zozobra worker, Zozobra is:

. . . a hideous but harmless fifty-foot boogeyman . . . He is a toothless, empty-headed façade. He has no guts and doesn't have a leg to stand on. He is full of sound and fury, signifying nothing. He never wins. He moans and groans, rolls his eyes, and twists his head. His mouth gapes and chomps. His arms flail about in frustration.

Crowd at Fort Marcy Park

Every year, we do him in. We string him up and burn him down in a blaze of fireworks. At last, he is gone, taking with him all our troubles for another whole year.

Zozobra Burning

Will Shuster turned Zozobra's reins over to the Kiwanis Club of Santa Fe in 1963. Since then, the Club has continued Shuster's tradition using the funds it generates to provide college scholarships for local high school seniors.

VIVA LA FIESTA!
VIVA ZOZOBRA!

DAYS OF THE DEAD

Location: Santa Fe Streets and Museums
Features: Parades, Exhibits, and Workshops
Contact: Museum of International Folk Art (505) 476-1200
or www.moifa.org

Dias de los Muertos (Days of the Dead) is a Mexican tradition that is not generally observed in New Mexico except, that is, in Santa Fe where nothing is like anyplace else.

Wise Fool Stilt-Walker at Day of the Dead Parade

The Aztecs believed in *el mas alla* (the afterlife) and set aside two days in late summer to honor *Miccailhuitontli* (the Small Dead) and *Hueymiccailhuitontl* (the Adult Dead). In preparation of these days, altars were set up, flowers were gathered, gravesites were decorated with gifts, and a variety of food was prepared. It was believed that the spirits of the departed loved ones resided peacefully in the afterlife but that, once a year, they returned home to visit family members. The flowers, food, and gifts were used to welcome them.

When Spain conquered Mexico in the 16th century, many of the native traditions were outlawed and replaced with those of Christianity. However, since the Aztec feasts honoring their dead were similar to the Catholic celebrations of All Saints' Day (November 1) and All Souls' Day (November 2), the Aztec tradition, slightly modified, merged with the Christian and became known as Los Dias de los Muertos.

For several centuries, the Days of the Dead were celebrated throughout Mexico. Mothers spent hours preparing *pan de muerto* (a sweet bread molded into the shape of a skull) and the departed one's favorite foods. Children went to the market to buy armfuls of yellow *cempazúchitl* (marigold-like flowers). Husbands repaired family altars and prepared their tools for the cemetery. When all was ready, the family journeyed to the cemetery, cleaned the gravesite, laid down the flowers, lit some candles, and sat down for an all-night vigil and picnic. It was a family reunion that brought everyone, alive or dead, together; it was a celebration of life after death. However, as the country became urbanized and people sought out better jobs, religious and cultural traditions became less important. By the 20th century, observance of Los Dias de los Muertos seemed all but extinguished.

In the 1960s and 70s there was a resurgence of Mexican cultural pride. Old traditions were revitalized, poems and books were written, movies were created, and exhibits were presented. An example of this process of cultural transformation was the recovery and re-invention of Los Dias de Los Muertos by artists groups and community art centers. Within a few short years, Days of the Dead celebrations began to appear almost everywhere. Los Dias de Los Muertos had returned.

Although its name might imply otherwise, New Mexico was settled by Spain and, since the Days of the Dead is a traditional Mexican holiday, it is not usually celebrated in New Mexico. However, since Santa Fe is also known as *The City Different*, it shouldn't come as a surprise to learn that this Mexican holiday found a home there.

Santa Fe's celebration of Dias de Muertos, also known as *Vivos Entre Los Muertos* (Living Among the Dead), is sponsored by El Museo Cultural de Santa Fe, the Museum of International Folk Art, the Hands-On Community Art Program, Wise Fool New Mexico, Warehouse 21, the Boys and Girls Club of Santa Fe and the Santa Fe Chapter of Veterans for Peace. Festivities, usually held the last two weeks of October, include parades, workshops, museum exhibits, live music, and lots of good food. Children, of all ages, learn to create *flores enceradas* (crepe paper flowers), *calaveras* (sugar skulls), and *papel picado* (cut paper banners). *Ofrendas* (altars) laden with food, candles, and photographs fill the museums, *Doña Sebastiana* flies by on her *carreta de la muerte* (death cart), giant puppets terrorize unsuspecting spectators, and calacas (humorous skeletons) cavort in unexpected places and unexpected manners.

Calacas

Most Mexicans view death as a transition between mortal life and eternal life. It is not something to be feared. On the contrary, most Mexicans welcome it. Nobel laureate Octavio Paz wrote:

The Mexican is familiar with death. He chases after it, mocks it, courts it, hugs it, sleeps with it, celebrates it. It is one of his toys and his most steadfast love.

Santa Fe's Days of the Dead celebrations may differ in form to those in Mexico, but the spirit of the tradition lives on confirming that life and death is at the core of human existence but also suggesting that some of us might be taking the whole thing far too seriously.

Wise Fool Ogre

SANTA FE NATIONAL CEMETERY

Location: 501 Guadalupe Street—Santa Fe
Established: Originally 1870, again in 1892
Features: Notable Burials
Contact: Cemetery Office / (505) 988-6400 or
www.cem.va.gov/nchp/santafe.htm

In 1870, the Catholic Diocese of Santa Fe donated land on the north side of the city for the Santa Fe National Cemetery, Six years later, the War Department decided it to too expensive to maintain the property as a National Cemetery and downgraded it to that of a post cemetery. After nine years of protest from families who thought their fallen heroes had been buried in a place of military honor, the cemetery was re-designated a national cemetery. Since that time, veterans from all the wars, including the Mexican-American War, the Spanish-American War, the Civil War, World Wars I and II, the Korean War, the Vietnam War, the Gulf Wars and Iraq have been buried in this cemetery. As of the end of 2003, 36,743 souls found their final rest here, several that were instrumental in shaping the future of New Mexico.

When the United States gained possession of New Mexico in 1846, Charles Bent of Taos became its first territorial governor. Although Mexico wanted to be the dominant power in North America, it was unable to achieve political control over the area. Consequently, the United States, in its "Manifest Destiny" movement west, stepped in. Even though the occupation was bloodless, not everyone was happy. Groups of insurgents, afraid that they might lose the land granted to them by the Spanish and Mexican governments, planned several uprisings aimed at recapturing New Mexico for the Mexican Republic.

On the night of January 18, 1847, a group of rebels aided by Pueblo Indians attacked Governor Bent at his home in Don Fernando de Taos. It was well known that Bent, like several other government officials, was sympathetic to the American cause. When the Governor opened the door, he was beaten, scalped, and left for dead. Bent's wife and children dragged him into the house but, unable to help him, watched as he died on the floor. Charles Bent's body was laid to rest in the Masonic Cemetery in Santa Fe where it remained until being moved, along with 49 soldiers, sailors, and government officials, to the National Cemetery in 1895.

Governor Bent Grave

New Mexico participated in the Civil War by sending volunteers to aid the Union troops at the battle of Glorieta Pass in 1862. The conquest of Glorieta Pass was essential to the Confederates because it would provide easy access to Fort Union, the Santa Fe Trail, and the gold mines of Colorado.

The battle began around noon on the 26th day of March. The Union soldiers, led by Major John M. Chivington, numbered around 400 while the (mostly Texan) Confederate soldiers, led by Colonel William Scurry numbered less than 300. Fighting went back and forth with one side attacking and the opposite side counterattacking. On the second day of fighting, reinforcements arrived increasing the Union troops to 1,200 and Confederates to 1,100. On the 28th of March, the Confederates succeeded in driving the Union troops from the Pass. Thinking they had won the battle, the Confederates returned to camp only to find that the Union soldiers had destroyed all of their supplies and pack animals, forcing them to retreat to Santa Fe and then back to San Antonio, Texas.

During the two-day battle, approximately 331 lives were lost. When the National Cemetery opened in 1870, the remains of 265 U.S. soldiers from the battlefields of Glorieta, Koslouskys, and Fort Marcy were moved there. In 1987, the remains of 31 Confederate soldiers, ranging in age from 17 to 42 and killed during the Battle of Glorieta Pass, were discovered in a mass grave while workers were excavating for a house foundation. After some debate as to where the bodies belonged, they were re-interred at the National Cemetery.

The Santa Fe National Cemetery is also the burial place of seven Medal of Honor recipients whose surnames reflect the rich diversity of New Mexico's heritage. They were:

Corporal Jacob Gunther. Indian Campaigns. Company E, 8th U.S. Calvary.1868/69.

Private Edwin L. Elwood (Chiricahua). Indian Campaigns. Company G, 8th U.S. Cavalry. October 20, 1869

Army Scout Sergeant Y.B. Rowdy (Yuma). Indian Campaigns. Company A, Indian Scouts. May 15, 1890.

Watertender Edward A. Clary. On board U.S.S. Hopkins. U.S. Navy. February 14, 1910.

First Lieutenant Alexander Bonnyman, Jr. World War II. U.S. Marine Corps. November 20/22, 1943.

Private First Class Jose F. Valdez. World War II. U.S. Army, Company B, 7th Infantry, 3rd Infantry Division. January 25, 1945.

Specialist Fourth Class Daniel D. Fernandez. Vietnam. U.S. Army, Company C, 1st Battalion, 5th Infantry, 25th Infantry Division. February 18, 1966.

A poem about the Mexican-American War is as relevant today as it was when it was written. One stanza reads:

Rest on embalmed and sainted dead!
Dear as the blood ye gave;
No impious footstep shall here tread
The herbage of your grave;
Nor shall your glory be forgot
While fame her records keeps,
Or Honor points the hallowed spot
Where Valor proudly sleeps.

From *Bivouac of the Dead*

by Theodore O'Hara—1854

Santa Fe National Cemetery

ROSARIO CHAPEL AND CEMETERY

Location: Paseo de Peralta at Guadalupe Street—Santa Fe
Established: 1807
Features: Rosario Chapel, Japanese Internee Graves
Contact: Cemetery Office / (505) 983-2322

The delicate wooden statue enshrined in the La Conquistadora Chapel in the Cathedral of St. Francis of Assisi was brought to Santa Fe by oxcart in 1626 by a Franciscan missionary, Fray Alonso de Benavides. When the Pueblo Revolt broke out in 1680, the statue was rescued by Josefa López Sambrano de Grijalva, a housewife, and carried to safety in El Paso del Norte. In 1691, Don Diego de Vargas made a vow to return the statue to Santa Fe and "first and foremost, personally build (a) church and holy temple, setting up in it before all else the patroness of the said kingdom and villa, who is the one that was saved from the fury of the savages."

Four months later a group of 800 colonists camped in an area outside of Santa Fe while De Vargas led his army into the city. A short battle ensued, the Indians were driven from the city, and the victors reclaimed their "Villa of Holy Faith."

In 1714, the battle-scarred Church of St. Francis of Assisi was rebuilt and a special alcove devoted to La Conquistadora was added. Not long after, parishioners began taking the statue out in procession from its chapel in the church to a site outside the city where colonists had camped upon arrival in Santa Fe. There, under a cottonwood and juniper branch shelter, La Conquistadora was enthroned for a nine-day novena in June. At the end of the nine days, another procession returned the statue to the St. Francis Church where it stayed until the following year when the procession was repeated. As devotion to La Conquistadora continued to grow, it became obvious that the old shelter of cotton and juniper branches had outgrown its usefulness so the people built a chapel on the spot where Don Diego and the colonists camped before reoccupying Santa Fe. They called it Rosario.

When Mariano Martinez became acting Governor in 1844, he designed a spacious alameda (park) in front of the chapel, planted trees, and dug an acequia (irrigation ditch) to carry water to the trees. But, as often happens with a change in administration, when the United States took over power in New Mexico, the acequia filled up with sand, the trees died, and the alameda reverted to the parched piece of desert it once was.

The tombstones in the Rosario Cemetery read like pages from a New Mexico history book. One section of the cemetery tells the stories of the priests and nuns that worked to bring Christianity to the state. Not far away are the graves of three sisters, all less than 2 years old, who died within a short period of one another.

New Mexico's past was difficult, but so is the present. In 2003, utility workers found the skeletons of 13 children and four adults in an abandoned cemetery in Santa Fe. Several months later, a funeral Mass was held and the bodies were re-interred at Rosario.

Sadly, there is a potter's field here where the remains of the forgotten or unknown are buried, without headstones. And, in a section of the cemetery known as the Rose Garden, two obscure and virtually forgotten tombstones tell the story of a world gone mad.

Following Japan's attack on Pearl Harbor on December 7, 1941, the FBI, under executive orders from Franklin Delano Roosevelt, identified and classified Japanese and Japanese-Americas based on criteria sent forth in a State Department report known as the Munson Report. The report stated that people of Japanese ancestry could be categorized either as *Issei* (first generation, immigrant), *Nisei* (second generation, born and educated in U.S.), *Kibei* (second generation, born in U.S., educated in Japan), and *Sansei* (third generation, infant or toddler). On February 19, 1942, more than 120,000 Japanese and Japanese-American men, women, and children, labeled as "dangerous enemy aliens," were forced to abandon their homes, friends, and businesses and, without trial, were imprisoned in relocation centers and internment camps. One of those camps was located in Santa Fe, across the road from the Rosario Chapel and Cemetery.

According to a monument placed on the grounds of the former internment camp:

4,555 men of Japanese ancestry were incarcerated (here) in a Department of Justice internment camp from March 1942 to April 1946. Most were excluded by law from becoming United States citizens and were removed primarily from the West Coast and Hawaii.

During World War II, their loyalty to the United States was questioned. Many of the men held here, without due process, were long time resident religious leaders, businessmen, teachers, fishermen, farmers, and others. No person of Japanese ancestry in the U.S. was ever charged or convicted of espionage throughout the course of the war . . .

Memorial at Frank Ortiz Park

The monument, a six-and-one-quarter ton granite boulder, sits upon a hill above the Frank Ortiz Dog Park in Santa Fe. When the idea for the monument was first proposed, it was met with a flurry of opposition from local VFW members. Most of those opposed to the monument felt that "honoring the internees was tantamount to slapping the faces of the vets." The October 1999 city council meeting at which the stone marker was approved nearly erupted in fisticuffs. After Santa Fe's Mayor Larry Delgado cast the tie-breaking vote, City Councilor Clarence Lithgow exploded in anger and shouted, "You just kicked the Bataan veterans in the teeth, in the twilight of their years."

Dedication ceremonies were held in Ortiz Park on April 20, 2002. Former internees, their families, and camp staff placed flowers at the base of the monument, exchanged stories, and shed tears.

In the 1950s, all traces of the camp disappeared when the Casa Solana subdivision was built. All that remains now are a granite boulder, two graves, and a lot of memories.

Tombstones of Japanese Internees at Rosario Cemetery

CHILDREN'S PEACE STATUE

Location: Ghost Ranch Santa Fe Campus, Paseo de Peralta and Old Taos Highway—Santa Fe
Established: 1995
Features: Bronze-cast Globe
Contact: NetWorks Productions / (505) 989-4482 or www.networkearth.org

At approximately 8:15 am on the morning of August 6, 1945, an atom bomb was dropped on Hiroshima, Japan. A huge fireball engulfed the city. Buildings crumbled, glass turned to liquid, and all combustible materials within three square miles disintegrated into dust. One-fourth of the city's residents died instantly; another fourth sustained critical burns or permanent loss of eyesight; thousands were forever changed by the black rain of radioactive poison that fell on the city.

SANTA FE NEW MEXICAN

Los Alamos Secret Disclosed by Truman

ATOMIC BOMBS DROP ON JAPAN

Deadliest Weapons in World's History Made In Santa Fe Vicinity

'Utter Destruction,' Promised in Potsdam Ultimatum, Unleashed; Power Equals 2,000 Superforts

More Nippon Cities Now Smoldering Ruins

The Santa Fe New Mexican Newspaper Headline

One such person was a twelve-year-old schoolgirl named Sadako Sasaki. One day, while she was out running with a friend, Sadako suddenly became dizzy. When she got home, she told her mother what happened. At first, the mother wasn't worried but when Sadako kept having dizzy spells, she took her daughter to a doctor. The doctor performed blood tests and discovered that Sadako had an illness called leukemia or what some Japanese people called the Atom Bomb Disease. The doctor explained that, although Sadako's disease had been dormant for ten years, it was now progressing rapidly.

Within days, Sadako was hospitalized. While there, one of her friends told her an ancient legend about a white crane that lived for 1000 years. According to the legend, the bird had mystical powers and anyone who folded 1000 origami cranes would be granted a wish. Sadako liked the story and decided to fold a thousand paper birds so that she might become well.

At first, Sadako enjoyed her task—it was fun and it kept her mind off her illness. Using whatever paper was available, she created white, blue, and yellow origami cranes. She painted faces and eyes on some of the cranes and drew pictures or wrote poems on the wings of others. One of the poems read, "I shall write peace upon your wings, and you will fly around the world so that children will no longer have to die this way." Eventually, however, it became difficult for Sadako to fold the paper. Her fingers hurt, her eyes blurred, and she grew weaker and weaker. Everyone wanted her to stop, but she never gave up. Before she died, Sadako completed 645 paper cranes. Her classmates finished the rest.

In memory of their friend and the other 7000 Japanese school children killed by the atomic bomb, Sadako's friends collected enough money to have a statue of a young girl holding a golden crane high above her head placed in the Peace Park in Hiroshima. At the base of the monument, thousands of folded paper cranes carry messages of peace from around the world. Every day, hundreds of students come to pray at the monument. An inscription in stone reads: "This is our cry: This is our prayer; Peace in the world."

In 1989, a group of 3rd, 4th, and 5th graders at Albuquerque's Arroyo del Oso School wanted to show how children could make a difference with regard to world peace so they decided to create some sort of "Peace" statue. A five-year "Dollar-A-Name" campaign collected 90,000 names of children from 50 states and 63 countries and paid for the construction of a bronze-cast suspended globe adorned with 3,000 bronze plants and animals created by children in over 100 countries.

Childrens' Peace Statue at Ghost Ranch Santa Fe Campus

The Peace Statue, dedicated in 1995, was intended as a gift from the children of the United States to the city of Los Alamos, the birthplace of the atom bomb. When the request for permanent placement there was brought before the Los Alamos City Council, the idea was tabled and the statue refused. For one year, it held a place of honor at the Albuquerque Museum and now is on loan to the Ghost Ranch Santa Fe Campus.

"Peace Day" is observed each year in Santa Fe on the 6th of August, also known as Hiroshima Day. On this day, people come together in the Santa Fe Plaza to discuss how and why peace is important. A white muslin peace crane circles the plaza, the gazebo is hung with long strings of colorful paper cranes, and children on the stage sing songs of love and world peace. At the end of the day, the paper cranes are taken to the Ghost Ranch Campus and hung on the Children's Peace Statue where they remain until the wind carries them away.

This is our cry:
This is our prayer;
Peace in the world.

Muslin Crane at Santa Fe Peace Day Celebration

STONEFRIDGE

Location: Paseo de Vista at Rincon de Torreon—Santa Fe
Established: 1996
Features: A Fridgehenge
Contact: Primordial Soup Company, Santa Fe

Stonefridge Sign

Hikers, bikers, and dog walkers may be the only people that know it's there, but it's there, nonetheless. What is it? Why, it's Stonefridge, of course and Stonefridge is a fridgehenge. Or is Fridgehenge a Stonehenge? Confused? Ok—let's back up a bit and start from the beginning.

It all began in southwestern England sometime during the Stone Age and Bronze Age, or between the years 3000 and 1000 BC. Someone, no one is actually sure who, decided to place some really big rocks in a circular pattern around some other rocks and a stone altar. Each of the rocks, or should we say stones, was neatly trimmed and topped with a lintel or top piece. Since no one is sure who arranged the stones, they also don't know how or why the stones were arranged as they were. Needless to say, there are theories. Some people think the stones were set in place by sun worshippers who had mystical powers. Others, can you believe it, say incredibly strong aliens from outer space moved the rocks around so they could have a landing space for their inter-planetary spaceships. Archaeologists, and other brainyacks maintain the stones tracked some sort of astronomical phenomena like eclipses and solstices while those who believe in witchcraft argue the circle of rocks was used for pagan gatherings and ceremonies. But, once again, no one is really sure. What is sure, however, is that somewhere along the way someone got the idea to replicate the "mysterious" rocks.

In 1979, an unidentified man (or men) in Georgia built a granite observatory. Known as the Georgia Guidestones, this Stonehenge wannabe included Astro-calendrical aspects and didactic "New Age" messages written in twelve different languages, warning onlookers to "avoid useless officials" and "unite humanity." Not to be outdone, students at the University of Missouri, the Show Me state, created a small-scale Stonehenge model (they called it the Missouri Megalith) with the added attractions of an analemma (a scale that indicates the declination of the sun and allows for the calculation of solar time) and a North Star viewing window. In Texas, a plaster and graphite-covered half-size version of Stonehenge was put together in someone's back yard. A couple of years later, an Easter-Island-like "head" was added. On the lighter side, Stanley Marsh III, also of Texas, up-ended 10 Cadillacs in the ground and called them "Cadillac Ranch." And, in Alliance, Nebraska, local residents gathered up a bunch of old cars, painted them gray, set them in a circle (lintels and all), and called them "Carhenge."

The most unique Stonehenge take-offs, however, are the Fridgehenges. The first was built in 1994 by a group of New Zealanders who erected a Stonehenge look-alike in a cow pasture. Made up of 41 old refrigerators, each corresponding to the original stones in England, the appliances were erected for a solstice celebration partly as a comment on the way that that occasion had changed over the past 4000 years and partly because it was a really good excuse for a party. Heckled by hymn-singing do-gooders and bolstered by a couple bottles of bubbly, the group battled the forces of nature (wind and rain) but finished their project at exactly 3:24 pm, the exact time at which the sun stood directly over the Tropic of Capricorn. Some rather impromptu solstice celebrations were followed by a free-form concert, of sorts, dedicated to the fridges. Later in the evening, solstice celebrants returned to the fridges, built a fire, sang some songs, and drank copious amounts of home-brew. According to David Riddell, one of the events organizers, "It was, all in all, a huge success." Never meant to be a permanent structure, the refrigerators were taken down to give way to the cows. A couple of years later in New Mexico the story was completely different.

In 1996, Santa Fe artist Adam Jonas Horowitz began construction on what he calls "a post-apocalyptic monument to waste and consumerism." Horowitz's plan called for stacking 200 white, avocado, harvest gold, copper tone, and almond refrigerators in an astronomically configured circle. Although scheduled for completion in 2003, the refrigerator megalith, surrounded by weeds and a rapidly deteriorating wire fence, remains unfinished because of bureaucratic red tape. Some people even argue that the old refrigerators present a safety hazard. Closer inspection, however, reveals that all of the refrigerators have been permanently sealed and power to the units was turned off a long time ago.

What a pity...wouldn't this be a great place to keep beer cold?

Fridgehenge

4

ALBUQUERQUE AREA

OLD ALBUQUERQUE

Location: At Central (Route 66) and Rio Grande Boulevard
Established: 1706
Features: San Felipe de Neri Church, Old Town Plaza and Gazebo, and Route 66
Contact: Albuquerque Convention & Visitors Bureau
(505) 842-9918 or or www.agqcvb.org
New Mexico Route 66 Association / (800) 545-2040
or www. rt66nm
Old Town Visitors Center / (505) 243-3215
San Felipe de Neri Parish Office / (505) 243-9708

In 1706, 252 settlers led by Governor Don Francisco Fernandez Cuervo y Valdez, made their way south from Bernalillo to the site of a new city they named La Villa de San Felipe de Neri de Albuquerque in honor of King Philip of Spain and Don Francisco de la Cuerva Enriques, the Duke of Albuquerque. As was customary in those days, the settlers laid out a town plaza, built their adobe homes around it, and began construction of a simple mission-style church on the west side of the Plaza. Taking more than 12 years to complete, the religious center included a church, a cemetery, a horse and mule corral, and a rectory. Following an unusually wet summer in 1792, the church collapsed and a new cross-shaped church was built on the north side of the Plaza. Constructed with five-foot thick walls, the new church was built to last.

Albuquerque grew as farmers and cattlemen discovered the fertile valley lands that surrounded it. By the 1850s, the population that once numbered less than 300 had burgeoned into more than 1,500 people. A new roof and twin bell towers were added to the church, shopkeepers set up their businesses around the Plaza, New Mexico became a United States Territory after the war with Mexico, and trade restrictions were lifted along El Camino Real. All in all, life was good. Then the Civil War broke out.

San Felipe de Neri Church

In 1861, Brigadier General Henry Hopkins Sibley, a former regular army officer organized a brigade of 3,200 Texas Confederates and planned an invasion of New Mexico, the gateway to the Colorado Territory goldmines. It was his intention to seize Fort Craig, Fort Marcy, and Fort Union, and then move into Colorado. Finding the forces at Fort Craig too strong to overpower, Sibley moved up the Rio Grande

River toward Fort Marcy. Hampered by snowstorms and feeling the distress of defeat, Sibley made it as far as Albuquerque where he planned to stock up on food and dwindling supplies. Unfortunately, the Union Army burned the supply depot and buried their cannon in the Plaza prior to his arrival. Nevertheless, Sibley raised the Confederate flag and stayed behind (in one of the best houses in town) while his men, then being led by Colonel William Scurry, advanced to Santa Fe, took over an abandoned Fort Marcy, and raised the Stars and Bars over the Territorial Capital.

During the spring of 1862, moving once again toward Fort Union, the Confederates were met at Glorieta Pass by a force of Union soldiers led by Major John M. Chivington. Fighting volleyed back and forth with first one side then the other gaining the upper hand. Finally, the Texans, slightly outnumbering the Union Troops, forced Chivington's men back. The Confederates had won the battle, or so they thought. Upon returning to their camp, Scurry and his men discovered that Union soldiers had destroyed his wagon train. The wagons contained everything the Confederates needed to continue their march to Fort Union. The Texans had no other choice than to retreat.

Back in Albuquerque, Scurry and Sibley received word that Colonel Edward Canby was marching toward the city with reinforcements from Fort Craig. A short battle erupted, several Confederates were killed (they are buried beneath the gazebo in the old Plaza), the remaining Confederates retreated to Texas, and the Union forces recaptured Albuquerque.

As the population of Albuquerque continued to grow, so did its need for space. In 1869, the San Felipe Church announced that the old camposanto near the original 1706 church was "too small and too frequented a place," and that the land would be used for a little plaza outside the church. Hundreds of bodies were carefully dug up and relocated to Santa Barbara Cemetery, a new cemetery that later became part of the Mt. Calvary Cemetery in New Town. Several years later, a second cemetery, a short distance from the church, was sold to local resident John Mann who planned to level the ground and use it to plant crops. Knowing the ground was once a cemetery, Mann agreed to let the church know if he turned up any bones. By the time the plot was leveled, more than two tons of bones were dug up. They were also reburied in the Santa Barbara Cemetery.

Plaque on Old Town Plaza

When the railroad came to Albuquerque in 1880, things began to change. After learning that the AT&SF planned to run tracks near the Plaza, local landowners priced their land so high that the railroad decided it would be less expensive if the tracks went through a less populated area two miles to the east. Local legend has it that when the depot station was built at this new location, the sign painter mistakenly left out the first "r" in Albuquerque, forever changing the town's name to Albuquerque. Fred Harvey built a small red lunchroom near the depot and called it "The Harvey House." A new town grew up around the depot and, although horse-drawn buggies shuttled people back and forth between the Plaza and the railroad station, all but a few die-hards deserted the old adobe houses around the Plaza in favor of the clapboard structures in "New Town." Old Albuquerque was dying.

As more and more people were attracted to Albuquerque, Fred Harvey decided it might be advantageous to tear down his old lunchroom and replace it with a full-fledged hotel. He hired California architect Charles Whittlesey to design the hotel, offered Mary Jane Colter the job of decorating it, and signed on dozens of "Harvey Girls" to tend to hotel visitors' wishes.

Harvey's Alvarado Hotel opened in 1902 and became an immediate success. It was an oasis in the desert and tourists arrived in droves. Many of them were actors, writers, and artists searching for the "Old West." They stayed in low ceilinged, simply decorated guest rooms, ate in Harvey's dining rooms, bought souvenirs in the Indian Building, and wrote long letters home while relaxing in the parlors. But the crisp, clean air, the azure-blue skies, the smell of roasting chiles, and the easy, laid-back lifestyle of Albuquerque were hard to forget. Many tourists stayed and became permanent residents, often in the nearly forgotten Old Town area.

There was new life in Old Town. Many of its old adobe houses, most in the state of near-collapse, were renovated, remodeled, and reborn. Some of the homes were transformed into shops and restaurants; others became workshops and galleries. Trees were planted in the Plaza, benches were put in, a gazebo was built, and the Church underwent extensive renovations. But there were even more changes ahead.

In 1926, Route 66 became the nation's principal east-west route, linking Chicago with Los Angeles. It was the way west. It was the Mother Road. It was unlike anything else. Cafes, service stations, and motor Courts sprang up along the road. Neon cowboys, cacti, and tepees enticed people to stop for food, gas, or the night. In Albuquerque, places like the El Vado Motel, the Moon Café, and the Jones Motor Company gained popularity while places like the Alvarado Hotel lost it. People were in a hurry to get wherever they were going and, in most cases, that wasn't

Albuquerque. The glamour of Hollywood and the glitz of Route 66 lured them on and, like lemmings, they followed.

Route 66 Sign

By the 1950s, air travel was increasing in popularity. Now, instead of driving all the way from Chicago to Los Angeles, people could hop a plane and be at their destination, usually in one day. Route 66 was by-passed. The Alvarado was by-passed. Albuquerque and its Old Town were by-passed. Neon signs went black; the Alvarado closed its doors in 1969; vandals took over the streets of Albuquerque. However, like children bored with a new toy, the nostalgia of the old road, the old hotels, and the old town eventually drew people back.

Recently, several restoration projects were begun to restore Albuquerque to its original grandeur. The New Mexico Route 66 Association is currently fixing or replacing many of its old neon signs; the City of Albuquerque is working to clean up and reclaim its streets; the Old Town Association has set up a visitor's center, offers free walking tours and has planned activities throughout the year; the San Felipe de Neri Parish has begun an ambitious restoration project to repair and remodel the exterior and interior of all the buildings in the church complex and a new transportation center, constructed to resemble the old Alvarado, has risen on the site of the original hotel.

The New Alvarado

Its appearance may have changed a little but the villa that grew up around the Plaza is still there . . . you just have to know where to look for it.

MOUNT CALVARY CEMETERY

Location: 1900 Edith NE—Albuquerque
Established: 1869
Features: Santa Barbara Cemetery, Woodmen of the World, and Gravel Pit Burials
Contact: Cemetery office / (505) 243-0218

At the request of Bishop Lamy, Jesuit priests from Italy came to Albuquerque in 1867 to take over church operations and oversee renovations of the buildings of San Felipe de Neri parish. Under the direction of Father Donato Gasparri, the old camposanto, rarely used in 15 years, was removed and replaced with a small plaza and convent. The exhumed bodies were taken to the new Santa Barbara Cemetery, three miles north of the church. Several years later, a second cemetery, a short distance from the church, was sold to local resident John Mann who planned to level the ground and use it to plant crops. Knowing the ground was once a cemetery, Mann agreed to let the church know if he turned up any bones. By the time the plot was leveled, more than two tons of bones were dug up. They were buried in a common grave at Santa Barbara.

In 1883, a man by the name of Joseph Cullen Root organized a society dedicated to "helping its fellowman." Based on a sermon Root heard one Sunday where a minister compared the community's need to work together to "pioneer woodsmen coming together and clearing forests to provide for their families," the society, initially named Modern Woodmen of American and later known as Woodmen of the World, became one of the first fraternal benefit societies in the United States. One of the benefits of belonging to this society was that, upon death, members would pass around a hat and collect money to provide a decent burial and gravestone for the departed member.

In the early 1900s, unique tree-trunk tombstones adorned with a dove, an axe, a mall, or a wedge, began appearing in the Santa Barbara Cemetery. Erected by the Woodmen of the World, the tombstone's sawed-off shape symbolized a life cut short and was often adorned with the tools of a woodsman. Although varying in size and shape, most of the tombstones bore the Society's Latin emblem, *Dum Tacet Clamat*, (Though silent, he speaks.)

WOW Tombstone

Although there are eight tree-like tombstones in the Santa Barbara and Mt. Calvary Cemeteries, placement of these once-popular vertical tombstones became rare and was eventually replaced by grand-level headstones and bronze markers, all reflecting the ideals and objectives of the Society that advocated them.

As the years went by and the population increased the need for a larger cemetery became evident. In 1936, 18 acres bordering the Santa Barbara Cemetery were purchased and the new Mt. Calvary Cemetery, which included a chapel, a mausoleum, and the old Santa Barbara Cemetery, was created.

1985, a mining corporation asked the University of New Mexico to conduct a cultural resources survey on land they wanted to use as a gravel quarry. The land was adjacent to the Mt. Calvary Cemetery and there was sufficient reason to believe it might contain some graves. As it turns out, more than 600 graves, both historic and prehistoric, were discovered in the course of the survey. Only 12 of graves could be identified. Some of the remains were re-interred at Mt. Calvary, the rest were reburied in a common grave at Fairview Cemetery.

There are many unusual and poignant gravesites at Mt. Calvary Cemetery but one, perhaps, is more heartbreaking than any other.

On the afternoon of May 10, 2000, while playing with a neighbor girl he was trying to impress, six-year-old Sergio Andrade soiled himself. In the haste of showering before his mother's boyfriend came home and discovered what happened, Sergio threw his dirty underwear in a pile of other clothes. When the 215-pound boyfriend came home, he discovered the underwear, grabbed a belt, and headed for the boy. After hitting Sergio several times with the belt, the man grabbed the boy by his shirt, threw him to the floor, and slammed his foot down on the boy's chest. Sergio's mother frantically searched for a phone to call the police but when she found it, the boyfriend attacked her, hit her in the head with the phone, and threw her to the floor. Then he returned to Sergio.

A Metro Court criminal complaint stated: the man "dragged the boy into the bathroom, began punching and kicking him, and slamming his little head hard enough into the wall that it left a hole." The boy went into convulsions and died a few minutes later. "Paramedics and emergency room doctors said the boy was covered in fresh bruises on his forehead, face, lower back, and both his legs. An autopsy report indicated that Sergio's liver had been split nearly in half." It was further learned that this was not the first time the boyfriend had beat the boy.

Sergio Andrade Grave

According to the National Network to End Domestic Violence, New Mexico is ranked the fifth worst state for domestic violence. Every year, innocent victims are forced to defend themselves against angry words, severe abuse, and vicious attacks. Numerous agencies and shelters have been set up to protect these victims but until the violence ends, we will continue to lose children like Sergio . . .

Dum Tacet Clamat,
Though silent, he speaks.

SAN JOSE, EL ROSARIO, AND BENINO CEMETERIES

Location: I-25 and Gibson off ramp—Albuquerque
Established: Late 1800s
Features: Fieldstone Markers and Wooden Crosses
Contact: Archdiocese of Santa Fe (Albuquerque)
(505) 831-8100

Back around the time the railroad arrived in the 1880s, the San Jose Cemetery was built on an isolated plot of land on the outskirts of Albuquerque. However, due to its remote location, few people were buried there.

In the summer of 1926, Route 66 became the nation's principal east-west route. Some ten years later, more than 210,000 people used it to escape the tragedy of the Dust Bowl. Driving dilapidated cars loaded down with their meager possessions, they made their way across the dehydrated prairie states in search of green fields, opportunity, and a better way of life in California. Some of them only made it as far as Albuquerque.

San Jose Cemetery Sign

The population of Albuquerque almost doubled during the 1920s-1930s. Although many people worked in the farm fields, others found employment with the railroad or service industries. For the most part, life was good. Modest neighborhoods sprung up, young men and women were married, children were born, people died, and a new cemetery, El Rosario was started south of the San Jose Cemetery.

For the next 40 years, loved ones were laid to rest at the San Jose and El Rosario Cemeteries, their graves outlined with field stones and their names scratched on homemade concrete tombstones or painted on wooden crosses. Then, in the 1960s, the land was divided into two parcels in order to make way for the construction of a new highway, I-25. San Jose and El Rosario Cemeteries were to the west of the highway; Benino Cemetery was to the east.

Over the years, drastic changes have come to these cemeteries. A cell tower was erected, weeds have grown up, the crosses have deteriorated, the concrete tombstones have eroded, and vandals have desecrated the graves. According to a recent survey, "fewer than 25 percent of the burials are recognizable . . . the cemetery gives the appearance of being abandoned."

How did this happen? Was it because the freeway was built or because vandals became a threat? Or was it because people simply lost interest?

Regardless of the reason, the harsh reality is that many cemeteries in the United States have been threatened, or destroyed, in the name of urban development, agriculture, logging, vandalism, and apathy. Although many states have laws protecting old cemeteries, few, if any, enforce the laws because it is too costly, too time consuming, and too controversial. And what about the families of the deceased? Shouldn't they take an active part in preserving their ancestors graves?

Not so long ago, family and friends performed all the activities associated with death. They built the coffins; they dug the graves; they buried their dead. At least once a year (birthdays, death dates, holidays) they returned to the gravesite, cleaned it up, planted some new flowers, and said a few prayers. However, as the funeral industry developed, the role of the family as caretaker changed. No longer having to build the coffins or dig the graves, once some family members witnessed the burial, they only returned to the cemetery when it was time to bury another loved one. As families moved away from caring for the graves of their deceased, hired personnel took over. Weed whackers, tractors, and industrial strength chemicals replaced hand tools and manual labor. The new equipment often caused damage to the tombstones and grave markers. Many cemeteries fell into disrepair and became eyesores that did not suitably reflect the quality of the surrounding neighborhood. Some of those cemeteries were destroyed.

Why does all of this matter? Because losing our cemeteries is like losing a part of our history. In a report on Cemetery Advocacy, Jeanne Robinson of the Oregon Historic Cemeteries Association stated:

> *(Cemeteries) are repositories of unique genealogical, historical, religious, cultural, societal, and medical information that may not be recorded in any other format...They are sources of humor, pathos, and folklore . . . They are places in which the average citizen has an opportunity to walk in the footsteps of their ancestors.*

Several years back, Guillermo Padilla, an Albuquerque resident, took it upon himself to clean up the San Jose, El Rosario, and Benino cemeteries. He painted the crosses, he picked up the trash, and he did whatever he could to help the cemeteries look respectable. Unfortunately, Señor Padilla died in 1999.

What does the future hold in store for these old cemeteries? Only time will tell.

Wood Cross in San Jose Cemetery

SUNSET MEMORIAL PARK

Location: 924 Menaul Blvd. NE (Edith and Menaul)
—Albuquerque
Established: 1929
Features: Rose Garden, Old Town Mosaics, and The Times and Seasons Columbarium
Contact: Cemetery Office / (505) 345-3536 or
www.sunset-memorial.com

As far back as 2500 B.C., cremation was readily accepted as a practical means of dealing with human corpses. Cremating the body ensured that animals and other humans would not desecrate it and guaranteed that the decomposing body would pose no threat to the living. However, early Christians considered cremation pagan and the Jewish culture favored traditional sepulcher entombment. As a result, by A.D. 400, earth burial almost completely replaced cremation.

At the turn of the century in 1900, only one percent of deaths in the United States involved cremation. In fact, so many people thought that it was such a strange and unusual practice that advocacy groups were formed to promote its use. Consisting mostly of doctors, scientists, and social reformers, the advocacy groups championed the cause of cremation, emphasizing its health and ecological benefits.

In 1913, Dr. Hugo Erichsen founded the Cremation Association of America and crematories soon sprung up across the U.S. In 1913, there were only 52 crematories in the United States; by 1975, there were 425; and in 1999, there were 1,468 crematories and 595,617 cremations.

Over the years, the acceptance of cremation has grown dramatically. In fact, more than 25 percent of deaths in the United States now involve cremation. By the year 2010, it is estimated that one out of every two people will choose to be cremated. Cemeteries like Sunset Memorial Park are preparing for that future.

Possibly the most beautiful cemetery in New Mexico, Sunset Memorial was built in 1929 as a place where families could go for comfort, beauty, hope, and inspiration. Set amidst the beauty of 75-year-old trees and spacious gardens, more than 32,000 kith and kin are interred at Sunset—some in the ground, others in permanent memorials.

Praying Hands at Sunset Memorial Cemetery

Instead of letting them drift away on the wind, some families scatter the ashes of their loved ones in a peaceful area known as the Rose Garden. Here, surrounded by marble, more than 50 floribunda rose bushes, eight trees, several benches, and winding paths, visitors sit, enjoy the view, and remember days gone by. Other families relax in the indoor comfort of the Times and Seasons Columbarium, a heated and air-conditioned bronze and glass structure with walls depicting each of the four seasons.

A short distance from the Rose Garden and the Time and Seasons Columbarium is the Old Town Mosaic, a columbarium constructed from 10,000 individual pieces of colored Italian glass. The artwork presents six scenes from around the Old Town area of Albuquerque including the San Felipe Neri Church, the church courtyard, the Sandia Mountains, and the Plaza Gazebo. It is divided into niches for burial urns and below each niche, the names of the interred are permanently inscribed.

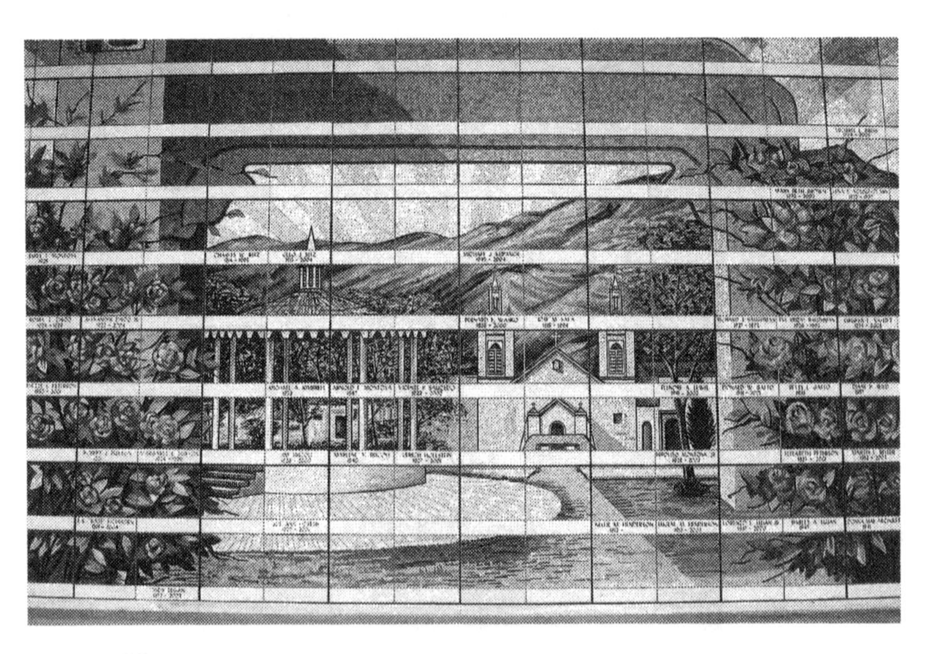

Old Town Mosaic

Losing a loved one is never easy. Neither is letting go. Some people say that, given time, we all forget but others believe we should never forget. In the book she coauthored with Jane Lester, South Australia health care therapist Barbara Wingard wrote:

Finding ways to bring people with us, those who are no longer living, can make a big difference in people's lives. When we reconnect with those we have lost, and the memories we have forgotten, then we become stronger. When we see ourselves through the loving eyes of those who have cared for us, our lives are easier to live.

Can places like Sunset Memorial help us reconnect with those who were lost as well as those who were left behind? Maybe so, maybe not. Either way, they make remembering a lot easier.

BARELAS BELL TOWER

Location: 412 Stover SW (Corner of 4th and Stover, south of downtown Albuquerque)
Established: 2003
Features: Salvaged World Trade Center Beams
Contact: Sacred Heart Catholic Church / (505) 242-0561

When Pedro Varela brought his family to the Albuquerque area in 1662, he had no idea what the future would bring. The land he settled was surrounded by swamps and marshlands infested with insects, reptiles, and unseen dangers. The arid climate and parched soil made farming difficult. But Varela didn't give up hope. He built a ranch, raised some sheep and cattle, and with the help of the Native Americans, planted a few crops. He also taught his children to put their faith and hope in God. His ranch prospered, his family grew, and more settlers moved in. Then the Pueblo Indians revolted against the Spanish Crown. Varela and his neighbors escaped to nearby El Paso del Norte where they remained isolated for twelve years. They prayed they could one day return to their homes and their prayers were answered when Don Diego de Vargas reestablished dominance over the Pueblo people in 1692.

When the railroad came through in the late 1800s, the population grew, the settlement's name was anglicized to "Barelas," and a church was built.

Dedicated to the Sacred Heart, the church had two bell towers, each housing a single bronze bell. The bells rang when World War II ended, they rang for weddings and funerals, and they rang every Sunday morning calling the people to worship. A rectory, school, and gymnasium were added to the church property in 1950. But the neighborhood was changing. In the 1950s, a sewage treatment plant was built in the area and the smell drove people away. The AT&SF roundhouse and rail yard, built in the 1930s, closed down in the early 1970s. Lack of money resulted in neglect of the church and it was razed due to structural problems.

The congregation moved services into the school gym but, somehow, the church's bells were lost. The barrio of Barelas was in decline—poverty, vandalism, drugs, and violence ravaged the once serene neighborhood. The situation seemed hopeless.

Several years ago, the people of Barelas resolved to take back their neighborhood. They renovated and reopened abandoned stores, set up shelters for the homeless, replaced graffiti with artistic murals and mosaics, and began construction of the National Hispanic Cultural Center to preserve, interpret, and showcase Hispanic arts and culture. One of the missing church bells was found in a parishioner's backyard and it was decided a bell tower would be added to the gymnasium so that the bell could once again ring signaling a new beginning for the neighborhood. Building the tower, however, was costly and residents didn't know where they would find the funds. There's an old Dicho (Hispanic saying) in New Mexico: "La esperanza es la ultima que muere—hope is the last thing that dies."

In response to the tragedy of 9-11, the people of New Mexico sent disaster assistance teams to Ground Zero. Not sure of what to do, the workers watched as bodies were pulled from the rubble, listened as sirens wailed through the streets, and felt the excruciating pain of families affected by the terrorist attacks. Everywhere, steel beams, once used in the construction of the towers, lay in twisted heaps. And then they knew . . . ship two of the beams back to New Mexico.

Rev. James Moore, now retired pastor of Sacred Heart Church, remembers: "Archbishop Michael Sheehan, House Representative Heather Wilson, and some others asked the City of New York for two of the beams from the ruins. We wanted to use the beams not only for a bell tower on the church but also as a permanent memorial to the people who perished in the disaster. At first, we were turned down but persistence paid off and our request was finally granted."

The beams were trucked in and, following a cross-country odyssey, arrived in New Mexico on May 25, 2002. Firefighters and law enforcement officers paid their respects by welcoming the steel beams with a full escort, an honor guard, and a 60-by-40 foot American flag that they draped over the beams.

WTC Beams

Sacred Heart Church Bell Tower

By June of 2003, the new bell tower at Sacred Heart Church in Albuquerque's historic Barelas neighborhood was complete. In a room beneath the tower, a plaque lists the names of the more than 2800 souls lost on September 11th. Pictures, paintings, even a quilt, bear testament to the loss.

Although only one of the church's original bells was ever found, it rings out today in memory of the people lost in the towers. It is dedicated to their lives—not to their deaths.

PETROGLYPH NATIONAL MONUMENT

Location: Unser Boulevard, Albuquerque West Mesa Area
Established: Prehistoric
Features: Ancient Petroglyphs
Contact: Park Office / (505) 899-0205

The black lava escarpment rising high above the streets of Albuquerque seems strange and out of place. And rightly so because it came from another time—a time before housing developments, automobiles, and jet planes—a time when mastodons, bison, and camel roamed the earth. So how did this anomaly get here and what does it mean to the people of New Mexico?

Long before man as we know him set foot on our soil, the earth's crust split and spewed several waves of molten basaltic rock across a 50 square mile area west of modern day Albuquerque. Volcanoes erupted, large areas of sedimentary rock were buried, and arroyos were filled. As time and erosion worked their magic, parts of the buried sedimentary rock eroded causing the overlaying volcanic rock to fracture and break into large boulders that formed the escarpment. Then, thousands of years of rain, mineral buildup, and microbe activity painted the boulders with a thin brown, or bluish black patina that scientists call desert varnish.

Two or three thousand years ago, groups of nomadic hunter-gatherers followed large herds of now prehistoric mammals into the area. They hunted, they gathered wild foods, they fashioned spear points and, even though the large game eventually disappeared, they adapted and stayed, many in the area of the escarpment.

Somewhere along the way, one of these people discovered that chipping away the dark patina on the rocks revealed a lighter color beneath. Within a relatively short time chiseled or pecked drawings of circles, spirals, hand prints, animals, stick people, ceremonial figures, and unidentifiable objects began appearing on the rocks. Almost 20,000 images were left on the rocks of the escarpment; most are on its east and south facing slopes.

Petroglyphs

Although their actual meanings may never be determined, archaeologists, ethnologists, and researchers have ascribed feasible interpretations to some of the identifiable petroglyphs. Abstract images such as wavy lines, circles, or spirals, may have been drawn to indicate water location, sun and moon cycles, or migration patterns. Animal representations most likely indicated the location of game or recorded a successful hunt. Kokopelli, the Johnny Appleseed of the Southwest, may have signified the beginning of agriculture. Deities were probably carved for many of the same reasons later people created statues and icons. Stories were told, important events were recorded, and prayers were expressed. The natives may have used the carvings to monitor the movements of the moon, stars, and plants or to guide them on a vision quest. The escarpment became a window to the past and a portal to the spirit world.

After Spanish Conquistadors arrived in the 16th century, life for the native people, now called Pueblo people, changed. The explorers brought livestock, friars, colonists, and guns. The Pueblo people, who up until then peacefully tilled the dry soil, prayed for rain, and lived off the land, were subjugated to a foreign lifestyle. They shared their food and homes with the explorers and helped them care for their animals. But something was wrong. The Spanish friars told the natives that their gods were bad and that only the Spanish gods were good. Spanish soldiers took most of the Indians' crops and colonists kidnapped their children and put them to work as slaves.

Even the rock drawings changed. Now, in addition to animals and abstracts, figures of men on horseback, herds of sheep, and Christian crosses began to appear. The native people rebelled and drove the Spaniards out in 1680. But the Spanish were persistent. Twelve years later they reentered New Mexico, reclaimed "their" land, and reconquered the natives.

One of the soldiers that helped recapture Albuquerque was Fernando Duran y Chaves, an early colonist. In recognition of his exemplary service in the Spanish army, he was awarded title to the Atrisco land grant. Part of that grant included the escarpment. Although it's not known whether or not any natives were living on the escarpment at the time, Chaves brought in livestock, introduced non-native plants to the grasslands, and permanently changed the once sacred land.

Some people say that the early inhabitants of New Mexico left no written history of their time here. Others believe that the petroglyphs were and still are a written history—one that cannot be read like ancient hieroglyphics but one, nonetheless, that recorded important events in the lives of the people that created them. As a history, it is important to preserve and protect the petroglyphs because they may be the only link to the Pueblo past. Once damaged, they can never be replaced.

In the1960s, concerned residents and students from the University of New Mexico's Geology Department joined forces to come up with a plan that would protect the petroglyphs. Their work led to the creation of the Indian Petroglyph State Park. In 1986, the Las Imagines National

Archeological District was included on the National Register of Historic Places. And, in 1990, Congress passed Public Law 101-313 designating the entire 7,300-acre escarpment a National Monument. Efforts to save the petroglyphs, however, were not over.

In recent years, there has been an on-going dispute concerning construction of a road through the Petroglyphs. People who live near the site contend they need easier access to and from their recently built homes. Pueblo leaders believe building a road through their ancient land would be sacrilege. William Weahkee, a Pueblo elder said: "Each of these rocks is alive, a keeper of a message left by the ancestors . . . There are spirits, guardians; there is medicine here."

Rocks at Petroglyph National Monument

Will the people of Albuquerque choose expansion over tradition? If they do, their decision could become yet another tragic chapter in the history of the Pueblo people.

SANTO NIÑO CAMPOSANTO

Location: Off Old Route 66, east of Albuquerque city limits in Carñuel
Established: Late 1800s
Features: Iron and wooden grave fences
Contact: Holy Child Parish / (505) 281-2297

Traffic on I-40 cuts through Central New Mexico at speeds well over 75 MPH and as the road approaches Albuquerque, it drops into a red rock canyon once known as Cañon de Carnué. The canyon, in use since prehistoric times, forms a perfect passageway between the Sandia and Manzano mountains and, although evidence of the region's earliest settlers is long gone, historians and visitors will find a small community, a church, and a cemetery.

Camposanto and Village at Carñuel

The present-day village of Carñuel, originally known as Carnué, began life with the San Miguel de Laredo-Carnuel Land Grant of 1763.

The request for this grant came from nineteen families who, despite the risk of attack by marauding Apache and Comanche Indians, wished to reclaim the lands their ancestors lost during the Pueblo Revolt of 1680. A plaza, or town square, was planned; houses were built; and agricultural fields were laid out. The settlers grew corn, wheat, apples, and peaches, raised some livestock, hunted rabbits and other small game, and did some trading with the Indians. The Indians, however, raided the settlers' fields, stole cattle, and terrorized hunters. In 1767, Santa Fe restricted further trade with the Indians and Albuquerque declared a moratorium on hunting. The settlers at Carnué were going hungry so they took matters into their own hands.

Some of the men began "finding" cattle in unusual places. One man came across an ox walking around the streets of Albuquerque and two others discovered some "dead" cows in the mountains. Although the men shared the newfound meat with their friends, dissension quickly arose among the other settlers. An investigation was held, neighbor brought testimony against neighbor, and even though the accused said it was "all a big mistake," arrests were made.

When a band of Apache Indians raided the plaza of Carnué sending everyone running for their lives, settlers wrote to the Governor and asked permission to abandon the land grant because of their "barbarous enemy." Naturally disappointed, the Governor tried to convince the settlers to stay but resistance ran high and all proposals intended to persuade them to stay were refused.

Twelve families, brave enough to withstand the continued threat of Indian attack, resettled Carnuel in 1819. By the 1890s, they built two chapels. A private chapel was built by Domingo Garcia to house the statue of San Miguel de Laredo, the village's patron saint, and a community chapel, Santo Niño, was built in 1898 on land donated by the Herrera family. Both of the chapels were torn down in the 1960s and a larger church was built.

The cemetery, built on top of an Anasazi village that dates back to A.D. 1100-1600 and located a short distance from the new church, features iron, wood, and wire grave fences. The fences defined the gravesite or family plot boundaries and served as protection against wandering cattle, coyotes, and other scavengers. While local blacksmiths crafted some of the original iron fences, others were purchased through mail order catalogs. Many of the iron fences disappeared in the scrap drives of World War II and were replaced by wood or wire fences that came from salvaged materials or feed and supply stores.

Fences at Santo Niño Camposanto

In 1932, Father Libertine erected a large white cross on the hill across the freeway from the church. The priest's idea was to encourage travelers through the canyon to stop and meditate. The cross is still there but many of today's travelers are in too much of a hurry to notice it . . . what a shame.

This lovingly cared for cemetery, hidden away from most eyes, seems to defy the onslaught of both time and progress. Looking much as it must have in the late 1800s, it stands as a testament to the value and importance of preserving traditions of the past. Hopefully, it will survive well into the next century.

RIO RANCHO VETERANS MEMORIAL PARK

Location: Pine and Southern Blvd, adjacent to Esther Bone Public Library in Rio Rancho
Established: 1996
Features: Veterans Memorial, Water Conservation Gardens
Contact: Rio Rancho Parks & Recreation / (505) 891-5015

Rio Rancho sprang to life in the 1960s after an ad was run in the Wall Street Journal promoting low-cost, low-interest homes located in a new development on the outskirts of Albuquerque. The ad appealed to Midwestern and Eastern residents interested in buying inexpensive property in an undeveloped area with a temperate climate and beautiful surroundings. In 1966, there were 100 families living in Rio Rancho. Ten years later the population swelled to 5,000, and in the 80s to 20,000. Today, more than 60,000 people reside in Rio Rancho.

Although the demography has changed somewhat over the years, most of the original settlers in Rio Rancho were retirees, many of whom served in the armed forces during the World Wars, Korean, or Vietnam. As new generations were born, grew up, and went out into the world, many followed their parents' example and joined the military. Within the first 30 years of Rio Rancho's existence, the United States was involved in seven different wars or police actions.

In the late 1990s, Rio Rancho's American Legion, Veterans of Foreign Wars, Disabled American Veterans, Jewish War Veterans, the Rio Rancho Honor Guard, and Purple Heart Veterans joined forces with the City of Rio Rancho and the Sandoval County Board of Commissioners to design and build a Veterans' Memorial to honor past and present Rio Rancho veterans. The Rio Rancho Veterans Memorial Park was officially dedicated in 1996.

Located adjacent to the city's public library, the Rio Rancho Veterans Memorial is unlike any other. Although there are four walls containing the names and branch of service of men and women who served in the military, the similarity to all other veterans' memorials ends there. In additional to the memorial itself, this park features a healing garden, a water conservation garden, a wetlands area, two gazebos, several picnic areas, and a multi-use sports field. Families and children come here not only to remember those who served our Nation, but also to watch birds, butterflies, and lizards, to learn about trees, native plants and water conservation (the Master Gardeners Club offers special classes and programs here), or to sit beneath a ramada and enjoy a good book. This is a place of joy and living but it is also a place for remembering. Those who need time to reminisce and contemplate can find peace and solitude while sitting on benches located around the monument and throughout the park.

The stark reality of war hits home when visitors enter the monument. A wall at the entrance displays commemorative plates listing the wars the United States has been involved in. They were the Revolution War (1771-1783), the War of 1812, the Mexican-American War (1846-1848), the Civil War (1861-1865), the Indian Wars, the Spanish-American War (April 1898-August 1898), World War I (1917-1918), World War II (1941-1945), the Korean War (1950-1953), the Vietnam War (1964-1975), Lebanon (1982-1984), Grenada (October 1983-December 1983), Panama (December 1989-January 1990), the Gulf War (1990-1991), Somalia (1992-1994), Bosnia (December 1995), Afghanistan (2002), and Operation Iraqi Freedom (March 2003—?) A few steps away, four two-sided walls contain terra cotta bricks listing the names (but not the ranks) of Rio Rancho veterans and people like the Blue Star Families and the Gold Star Wives who supported them. The four walls are almost full but two empty walls stand ready for additional names. Several obelisks detail the number of those killed, wounded, or declared missing during some of the more recent wars. During World War I, 116,516 men and women were killed, 204,002 were wounded, and 4,500 were missing. During World War II, 407,318 were killed, 671,801 were wounded, and 8,714 were missing. During the Korean War, 54,246 were killed, 103,284 were wounded, and 5.178 were missing. During Vietnam, 58,167 were killed, 153,303 were wounded, and 2,325 were missing. And, during the 1990-1991 Gulf War, 147 were killed, 458 were wounded, and 23 became prisoners of war. So far, casualties from the more recent wars have not been inscribed on park obelisks.

Sculpture at Rio Rancho Veterans Memorial Park

A bronze plaque expresses the sentiments of the city:

For over two centuries, American Veterans have pledged their lives, their fortunes, and their sacred honor in defense of this Great Nation. A grateful city dedicates this monument to the men and women of our armed forces, whose courage, idealism, and sacrifices perpetuate American Freedom.

Another plaque expresses the sentiment of the veterans:

We served so America can be Free.

Memorial Wall

SAN YSIDRO CAMPOSANTO

Location: Old Church Road in Corrales
Established: 1868
Features: Rescued Caskets, Old Tombstones, and Ghosts
Contact: Archdiocese of Santa Fe (Albuquerque)
(505) 831-8100

In 1710, Francisco Montes y Vigil, a corporal in the Spanish Army, received a Royal grant to an "uncultivated and unsettled" land tract bounded "on the north (by) the ruin of a pueblo, on the south (by) a small hill, on the east (by) the Rio del Norte (Rio Grande), and on the west (by) prairies and hills." Known as the Alameda Land Grant, the property may have originally contained as much as 100,000 fertile acres of land. Two years later when Vigil determined that he could not occupy the land, he conveyed the grant to Captain Juan Gonzalez who built a hacienda for his family and a riverfront church for the workers who lived on his land.

By the mid 1700s, there were about 200 people living in Corrales. One hundred years later, the population had grown to a little over 600. Many of the settlers worked for the large land-holding patrones but some owned small farms or ranches of their own. Almost all of the settlers raised sheep and cattle; others grew alfalfa, chiles, or grapes. The people tended their livestock, tilled their fields, raised their families, and attended Mass every Sunday at the church near the river. For the most part, life in Corrales was peaceful. Then disaster struck.

During flood season, the Rio Grande often rose above its banks and inundated the surrounding countryside ravaging anything in its way. Although the details are sketchy, it is probable that such was the case in 1868 when the river flooded, literally changed course, and destroyed the Corrales church. The July 1st edition of the Santa Fe Weekly New Mexican reported the incident:

At Corrales, the camposanto with all the remains, and a number of buildings, have been washed away, among them the church.

Historic San Ysidro Church

The people of Corrales "donated their labor, any building materials they could spare, and even their hard earned pennies" and built a new church, dedicated to San Ysidro, the patron saint of farmers. Located away from the river, the church included a small plaza surrounded by a low wall and a new camposanto built on land donated by one of the local patrones. The first burials in the cemetery included caskets recovered from beneath the floor of the old church and its waterlogged graveyard.

The new camposanto was set up like any other camposanto of the time. The graves were close together, most of the tombstones were hand-made, and wood or iron fences enclosed family plots. The graves of Catholics had their feet pointing toward the east so that the "departed" would be ready to rise up and face the new day when the "trumpet shall sound and the dead shall be raised." Those of non-Catholics, servants, suicides, or unknowns were buried with their heads to the north

San Ysidro Camposanto

Over the years, many of the oldest graves and markers succumbed to the elements. Rain dislodged and washed away many of the rocks; the sun baked and dried out wooden crosses and fences causing leaving them to fall away in splinters; snow and ice seeped into and eroded old cement headstones causing them to collapse. Some of the tombstones were lost forever but a few endured the onslaught of time and survived into the 21st century.

At the east entrance to the camposanto, a 5-foot-tall white marble statue of a woman with her arms around a cross towers above the grave of Ignacio Gonzalez. Ignacio was a wealthy descendant of Captain Juan Gonzales who bought the Alameda land grant from Francisco Montes y Vigil. Although the epitaph on the pedestal supporting the statue is hardly decipherable, the date of 1896 remains legible. Nearby, the graves of infants and children are gathered together—the dates on most of their tombstones painfully clear. It's comforting to think someone has been caring for them.

Statue at San Ysidro Camposanto

Another tombstone, made almost entirely of beer bottles, reflects the humor sometimes evident in graveyards . . . the person who drank the beer didn't survive but the bottles he drank from obviously did. Of all the graves in the camposanto, however, two stand out because of their absence—those of Luis and Louisa Emberto.

Leaving France in the late 1800s, Luis and Louisa purchased an almost 100-year-old hacienda in Corrales and settled down to what they hoped would be a pleasurable life. Part of that life included extravagant soirées and fiestas held at their hacienda. It was said that Louisa was a woman of great beauty and that although Luis had long suspected her of infidelity, he had no proof until the night he discovered her with another man. Overcome with jealousy, Luis stormed out of the hacienda promising to return and kill Louisa and her lover. On April 30th of 1898, he made good on his promise.

Using the dark as his cover, Luis quietly slipped into the hacienda and discovered Louisa and her lover in the great salon. He shot Louisa twice as she ran for her gun. Then he ran from the hacienda.

A posse quickly formed and pursued Luis across the fields and back roads of Corrales. When they caught up with him, a one-eyed Indian sharpshooter named Jose de la Cruz aimed true and killed Luis.

Due to the unholy circumstances surrounding their deaths, the Embertos were not buried in the church camposanto. Instead, they were placed in unconsecrated ground near an irrigation ditch west of the hacienda (now a restaurant). For many years, local residents reported sighting the ghosts of Luis and Louisa lurking around the property.

The village of Corrales has seen dramatic change since the 1800s. The once great land tracts have been subdivided into small farms and residential areas, the large herds of sheep and cattle have been replaced by goats and horses, many of the old homes have been transformed into shops and restaurants, and the church that was built after the 1868 flood is now a cultural center. And what about ghosts? Do they still haunt the village?

One can only wonder . . .

EL CERRO DE TOMÉ AND PUERTA DEL SOL
(The Hill at Tomé)

Location: West of NM 47 on NM263, approximately 30 miles south of Albuquerque in Tomé
Established: Prehistoric—Area settled in 1650
Features: Petroglyphs, crosses on hilltop, La Puerta del Sol sculpture
Contact: Greater Belen Chamber of Commerce
(505) 864-8091

Surrounded by small farms, mobile homes, and a few strip malls, the hill at Tomé, El Cerro de Tomé, seems uninviting. Rising almost four hundred out of the flood plain along the Rio Grande, there are no trees and little vegetation to soften its appearance; its rocky slopes make hiking hazardous; menaces such as snakes and scorpions hide everywhere. It isn't a likely candidate for New Mexico's most appealing attraction. Yet, for more than 2000 years, it has served as a gathering spot, an observation point, and a source of spiritual renewal. What is it that draws people to this austere hill?

The hill came into being millions of years ago when volcanoes played their role in shaping the familiar landscape of today's New Mexico. A sifting of tectonic plates created a rift that soon filled with water and sediment, creating the Rio Grande or Big River. Camels, ground sloths, and dinosaurs drank at the river's shores and searched the hill's surface for prey.

Some time later, no one is sure exactly when, humans moved into the area. They hunted the animals and etched images and designs on the hill's rocky outcroppings. The etchings, known as petroglyphs, represented birds, animals, fish, mythical creatures, and geometric patterns. The people that created the petroglyphs, probably the Anasazi, eventually disappeared leaving no written explanation of their work. The Pueblo People who followed respected the drawings of their ancestors and often left prayer offerings near the petroglyphs.

In 1650, Tomé Dominguez arrived in the area, set up a small estancia (homestead), and imparted his name to a hopeful, new community. His hopes were dashed, however, when the Pueblo Revolt of 1680 drove him and the other few settlers out of New Mexico. The settlement lay abandoned until July of 1739 when several families from Albuquerque reoccupied the area, created a plaza, and built a church. The church, named Nuestra Señora de la Concepción de Tomé, became the spiritual center of the community. Its three-foot-thick walls provided refuge during numerous hostile Indian raids. Because El Cerro (the hill) was an easily identifiable landmark, it established the settlement as a major stopping point along El Camino Real, the Royal Road between Old Mexico and Santa Fe.

One of the problems settlers encountered by living in rural areas was the lack of spiritual leaders. To overcome this loss, they formed a lay group known as Los Hermanos Penitentes and practiced a folk religion that emphasized penance and redemption. At Nuestra Señora de la Concepción de Tomé, El Cerro became the Calvario, or crucifixion site, where most Penitente processions, particularly those enacted during Easter, culminated. However, the ceremonies of the Penitentes came under scrutiny and derision when Bishop Jean Baptiste Lamy arrived in Santa Fe in 1851. From Lamy's point of view, the settlers' religious practices were bizarre and idolatrous and had to be stopped. The vexing actions of the Penitentes became shrouded in secrecy and processions up the hill all but disappeared.

By the late 1940's, resistance against Penitente practices had lessened and in 1947 when Edwin Berry returned to his home near Tomé after World War II, he and two Penitente group members raised three crosses on the top of the hill. The crosses represented a tribute to Berry's father, a long time Penitente member and group leader, and a return to traditional ways. Once again, pilgrims, penitents, and hikers climbed to the top of El Cerro for both the effort and the experience.

In 1997, a steel sculpture featuring life-sized native dancers, conquistadors, friars, sheepherders, railroaders, and a man in a sombre

Crosses on Top of Tomé Hill

ro—all marching through a 25-foot-tall arch—was erected at the base of El Cerro. New Mexico artist Armando Alvarez created the sculpture, La Puerta del Sol (Gateway to the Sun) to represent the diverse groups of people who traveled El Camino Real throughout New Mexico's turbulent history. Like the petroglyphs on the rocks and the crosses on the top of the hill, the sculpture stands as a monument to the many cultures that built the community and climbed or journeyed past this hill.

Today, the hill at Tomé serves many purposes. During the annual Good Friday Pilgrimage, people walk, run, or crawl up El Cerro. Some sing as they walk; others pray. Some carry wooden crosses; others put rocks in their shoes. Saturdays and Sundays find amateur archaeologists

Good Friday Pilgrimage

looking for petroglyphs, rock hounds searching for unusual specimens, and dirt-bikers practicing their skills on the loose rocks. Year around, sightseers come to see the sculpture, part of a statewide project called "Cultural Corridors: Public Art on Scenic Highways," and El Cerro—the only natural feature on the National Register of Historic Places. The hill is never deserted—there is always someone there.

La Puerta del Sol Sculpture

So, what is it that draws people to El Cerro? Is it the hill's archeological history, its religious significance, or its stark beauty? Is it for penance or exercise? Is it the desire to know something about the past or the future? Maybe the only way to answer these questions is to climb the hill and decide for yourself.

5

SOUTHERN NEW MEXICO

BILLY THE KID'S TOMBSTONE

Location: Old Fort Sumner Museum, US 60 and Billy the Kid Road, 7 miles SE of Fort Sumner
Established: Sometime around 1882
Features: Graves of William H. Bonney and a few of his friends
Contact: Old Fort Sumner Museum / (505) 355-2942

You might get an argument from some Texans but most people believe Henry McCarty, AKA William H. Bonney, William Antrim, Henry Antrim, Kid Antrim, and—of course—Billy the Kid, was shot, killed, and buried in New Mexico.

Roadside Sign

It all began in 1880 when New Mexico's Governor Lew Wallace put a $500 reward on Billy's head for failure to appear in court to stand trial for the alleged murder of Andrew "Buckshot" Roberts. Billy was well known for theft and cattle rustling and long-suspected, but never convicted, of numerous murders. Unfortunately, Billy moved so fast that lawmen were never quick enough to take him into custody. Wallace believed his reward would accomplish what he could not and that it would bring Billy to his knees and justice to the Southwest.

Sheriff Pat Garrett, a one-time friend intent on bringing his buddy in, formed a posse that pursued Billy and his gang throughout southern New Mexico. Garrett and his men killed Billy's cohorts in crime Tom O'Folliard in Fort Sumner, and Charlie Bowdre near Stinking Springs where Billy and the gang were holed up. Stinking Springs, as the name might imply, was not a very hospitable place. There was no water, food supplies were dangerously low, and Garrett was picking off gang members one by one. So Billy surrendered.

Garrett took Billy to Mesilla (near present day Las Cruces), a trial was held, and Billy was convicted of the murder and sentenced to be hanged. However, before the execution could take place, Billy escaped, killing two of Garrett's deputies in the process.

Three months later, Garrett gunned down 21-year-old William Bonney in a darkened cabin on Pete Maxwell's ranch in Fort Sumner where the two men once worked. Billy's body was buried on the Fort Sumner ranch but treasure seekers forced authorities to move his remains to the Old Fort Sumner Military Cemetery where he was re-interred with his "pals" Tom O'Folliard and Charlie Bowdre. In 1908, floodwaters from the Pecos River leveled what remained of the old fort, taking the house where Billy was killed with it. All that was left was the tombstone.

In 1950, Billy's tombstone was stolen from the cemetery. For more than 25 years, no one knew its whereabouts. It was finally tracked down in Granbury, Texas and returned to Fort Sumner, only to be stolen again and taken to California. When it was finally retrieved, caretakers at the Old Fort Sumner Museum decided it would be prudent to place it within an iron enclosure, away from marauders and vandals. Somehow, people still manage to get flowers, wreaths, and moments inside the three-sided enclosure. And, of course, there are pennies, nickels, and dimes everywhere.

Billy the Kid's Tombstone

Now here's where the disagreement comes in. Some people believe that Billy had yet another alias, Ollie L. "Brushy Bill" Roberts and that he actually escaped Garrett's bullets, hid out in Mexico, and died of old age in Hico, Texas. If that is true, there are two problems: someone other than William H. Bonney was buried in the Fort Sumner cemetery, and Sheriff Pat Garrett murdered that person.

In 2003, a Lincoln County sheriff, working with the State of New Mexico, proposed an investigation to determine whether Billy was buried in New Mexico or Texas. Not only would such an investigation end a long-time feud between neighboring states, it would also resolve the issue as to whether or not Pat Garrett shot the wrong man. The plan was to compare the DNA of Catherine Antrim (Billy's mother) to that of "Brushy Bill." If a match was made then good ol' Brushy and the Kid were, in fact, one and the same. If no match was made, then it might safely be assumed that the body once buried in New Mexico was that of William H. Bonney. However, for one reason or another, the investigation keeps getting delayed.

Who was the "Real" Billy the Kid and where was he buried? We may never know because, sometimes, it's just not possible to separate legend from reality.

PAT GARRETT'S GRAVE

Location: Masonic Cemetery, Las Cruces
Established: 1908
Features: Lawman's Grave
Contact: Las Cruces Convention & Visitor's Bureau
(505) 541-2441 or www.lascrucescvb.org

Born in Alabama and raised in Louisiana, nineteen-year-old Patrick Garrett came west in 1869 to hunt buffalo in Texas. When the herds thinned out and competition became fierce, Garrett got into an argument over some hides with another hunter, shot and killed the man, and then high-tailed it to New Mexico where he went to work on a ranch owned by Pete Maxwell. That's where he first met Billy the Kid.

Garrett knew that Billy, as one of John Tunstall's loyal Regulators, had taken part in the infamous Lincoln County War but, since he himself had not taken part in the battle, he didn't think Billy's escapades were any of his concern. However, things started to change in 1880 when Garrett was elected sheriff of Lincoln County.

Vowing to bring the rampant lawlessness in Lincoln County to an end, Garrett formed a posse to capture Billy, alive or dead, and collect the $500 reward Governor Lew Wallace put on Billy's head for failure to appear in court for the alleged murder of Andrew "Buckshot" Roberts. Garrett and his men tracked Billy and his gang throughout southern New Mexico, trapping him and some of his friends in a one-room house near Stinking Springs, New Mexico. Two of Billy's friends were shot, the rest were shackled and taken by buckboard to Las Vegas, New Mexico to await a train that would take them to the state prison in Santa Fe. Garrett took Billy to Mesilla (near present day Las Cruces), a trial was held, and Billy was convicted of the Roberts murder and was sentenced to be hanged. However, before the execution could take place, Billy escaped, killing two of Garrett's deputies in the process.

REWARD

($5,000.00)

Reward for the capture, dead or alive, of one Wm. Bonney, better known as

"BILLY THE KID"

Age, 18. Height, 5 feet, 3 inches. Weight, 125 lbs. Light hair, blue eyes and even features. He is the leader of the worst band of desperadoes the Territory has ever had to deal with. The above reward will be paid for his capture or positive proof of his death.

PAT GARRETT, Sheriff.

DEAD OR ALIVE!
"BILLY THE KID"

Replica of Billy the Kid Wanted Poster

Three months later, Garrett paid a visit to his old employer, Pete Maxwell, thinking that he might have some information as to the whereabouts of Billy. While the two men sat talking in Maxwell's darkened cabin, Billy walked in, became aware that someone was in the cabin, cocked his revolver, and hoarsely whispered "Quien es?" (Who is it?) Garrett fired twice, one bullet striking Billy squarely in the heart, the other ricocheting wildly around the room.

Billy had achieved a rather legendary reputation as one of the West's greatest scoundrels so, instead of acclaiming Garrett as the hero that brought "The Kid" in, the townsfolk scorned him for killing one of their favorite "sons."

Garrett was not reelected as Sheriff of Lincoln County, he lost an election for state senator in 1884, and word has it that Governor Wallace never paid him the $500 reward. Following an unsuccessful stint in the Texas Rangers, Garrett moved to Roswell, where he ran for sheriff of Chaves County. Needless to say, he lost that election, as well.

In October of 1899, Pat Garrett was appointed sheriff of Dona Ana County where he took on an investigation into a heinous crime involving the murder of Colonel Albert Jennings Fountain and his eight-year-old son, Henry. It took almost two years, but Garrett apprehended two suspects, Oliver Lee and Jim Gililland, charged them with murder, and brought them to trial. Both men were acquitted but things seemed to be looking up for Garrett.

On December 20, 1901, shortly after taking office, President Theodore Roosevelt appointed Garrett to the position of El Paso Collector of Customs. Garrett and his family moved to El Paso where he met Tom Powers, a reprobate run out of his home state for beating his own father into a coma. Aside from drinking and gambling, the two men had very little in common. Nevertheless, they went everywhere together, including to a reunion of Roosevelt's old combat unit, the Rough Riders. Roosevelt was appalled when Garrett allowed Powers to sit at the president's table and appear in a photograph taken of the group. Garrett lost his appointment in El Paso, returned to Dona Ana County, and moved to a ranch in Bear Canyon, about a four-hour ride from Las Cruces. He bought some horses and tried breeding quarter horses. His efforts went unrewarded.

Jobless and with a large family to feed (he and his wife had nine children) Garrett decided to sell the ranch. Meanwhile, in order to meet some of his day-to-day expenses, he leased some pastureland to Wayne Brazel, a cowboy who worked on the nearby ranch of W.W.Cox. What Brazel didn't tell Garrett was that he intended to graze goats on the land, not cattle. Garrett knew goats would bring down the value of the ranch and run off the few prospective buyers. He tried to break Brazel's lease but Brazel objected, stating that anyone interested in buying Garrett's land would have to buy his 1800 goats as well. Carl Adamson, a cowboy from Texas, said he would buy the Bear Canyon Ranch but not the goats. Garrett said he would see what he could do.

On February 29, 1908, Garrett and Adamson were headed toward Las Cruces when they came upon Brazel. An argument broke out and Garrett was shot once in the back of the head and once in the belly. Adamson and Brazel rode on to Las Cruces, leaving Garrett's body where it lay, and both claiming self-defense when they reached town.

Some people said that Adamson and Brazel worked together to kill Garrett. Others believe that Cox, upset with Garrett over some disputed water rights, used his employee, Wayne Brazel, to aggravate Garrett and run him off the land and that if Brazel couldn't get the job done, Jim Miller, a hired gunman could.

Brazel was brought to trial on a charge of first degree murder but was subsequently acquitted; Adamson died of typhoid fever; Cox got the water rights he wanted when he bought out Garrett's widow; Miller was apprehended and hanged in Oklahoma; and Garrett was buried in the Las Cruces Cemetery.

No one is sure who killed Patrick Floyd Jarvis Garrett but most people agree that he died the way he lived—tragically.

SMOKEY BEAR HISTORICAL STATE PARK

Location: Highway 380 (Smokey Bear Blvd.) in Capitan
Established: 1976/79
Features: Museum and Smokey Bear's Burial Site
Contact: Park Office / (505) 354-2748

The Capitan Mountains in New Mexico's Lincoln National Forest are one of the few mountain ranges in the country that run east to west. Varying in height from 5,400 feet to more than 10,000 feet, the peaks and valleys of these mountains pass through five different ecological life zones and are covered in dense stands of mesquite, yucca, and cactus in the lower elevations and piñon pine, fir, juniper, and spruce trees in the higher. Mule deer, coyote, wild turkey, golden eagle, and black bear call this region home. Abundant moisture in late summer and winter keeps the vegetation luxuriant but the high winds of spring parch the forest and aggravate the likelihood of fire.

In 1950, a devastating wildfire, reportedly caused by a carelessly discarded cigarette, swept through the forest charring trees and killing wildlife over an area of 17,000 acres. During the blaze, several firefighters noticed a small bear wandering along the fire line near the small town of Capitan. The firefighters assumed the cub had become separated from his mother and that, sooner or later, the mother bear would find him. But that never happened. Several days later the cub was found, badly burned and near death, clinging to the trunk of a blackened tree.

The tiny bear was rushed to a veterinary hospital in Santa Fe and then spent several months recuperating in the home of a game warden. When he was found the cub was named "Hot Foot Teddy," but later rechristened to "Smokey Bear" giving life to the animated bear that began appearing on forest fire prevention posters several years earlier.

As happens with all babies . . . the cub grew up. So much so, that he could no longer live in the game warden's home. Smokey was flown to the National Zoological Park in Washington, D.C. where he lived for 27 years. With only donated materials and local labor, the town of Capitan built a log museum to tell the story of the cub's traumatic start in life.

Over the years, millions of admirers visited the National Zoo to see Smokey Bear, the "Living Symbol" of the Forest Service's campaign to prevent forest fires. When he died in 1976, Smokey was returned to Capitan and buried in a garden next to the log museum.

Smokey Bear's Grave

In later years, the Smokey Bear Museum was enlarged and designated as the Smokey Bear State Historical Park. The visitor's center at the park includes exhibits about forest fires, a history of the fire prevention campaign, and more than 500 pieces of Smokey memorabilia including the original baby bottle used to feed the cub when he was rescued. An outdoor exhibit features plants from each of the six life zones found in New Mexico, an amphitheater that is used for educational programs, and the final resting place of the living symbol himself—Smokey Bear.

Fire has always played a vital and beneficial role in maintaining healthy forests. It promotes vegetative and wildlife diversity, helps maintain wilderness and wild land areas, reduces the population of disease-carrying insects, and eliminates excessive fuel accumulations that can ultimately lead to catastrophic wildfires. Every year, more than 4 million acres are consumed by wildfire. It is estimated that only 10% are from natural causes; the other 90% are caused by human error. Can anything be done to change these odds?

Yes—the Smokey Bear website (www.smokeybear.com) recommends the following:

- If smoking is permitted outdoors, safe practice requires a 3-foot clearing around the smoker.
- Never throw cigarettes or ashes out a car window
- Don't park vehicles on dry grass.
- Travel only on trails and other durable surfaces.
- If off-road vehicle use is allowed, internal combustion equip ment requires a spark arrester.
- Know the area's outdoor burning regulations. Unlawful burning is a punishable offense.
- Keep campfires small
- Never take burning sticks out of a fire.
- Never leave a campfire unattended
- Never take any type of fireworks on public lands.
- Keep stoves, lanterns, and heaters away from combustibles.
- Store flammable liquid containers in a safe place.
- Never use stoves, lanterns, and heaters inside a tent.
- Inspect campsite before leaving.
- At the first sign of a wildfire, leave area immediately by estab lished trails or roads. Contact a Ranger as soon as possible. If escape route is blocked, go to the nearest lake or stream.

And above all, remember . . .ONLY YOU can prevent wild fires!

Forest Service Sign

HAM THE ASTROCHIMP MEMORIAL

Location: New Mexico Museum of Space History, Alamogordo
Established: 1983
Features: Museum and Astro Chimp Grave
Contact: Museum Office / (505) 437-2840 or www.spacefame.org

Six chimpanzees were chosen by NASA to take part in the Mercury Program, America's first manned space adventure. Chosen because their reaction time, organ placement, and internal suspension was similar to that of man's, the primates were trained to perform several tasks, such as pulling a lever in response to lights and sounds, and sitting for extended periods in one seat. The reward for correct responses was banana pellets; the penalty for incorrect behavior was a mild electric shock to the soles of their feet. After completing their "Basic Training" at Holloman Air Force Base near Alamogordo, the chimps were transferred to Florida where they received additional training in a more humid, lower altitude environment. Scientists wanted to be sure that any biological changes that occurred in space were an effect of being weightless and not due to the stress of traveling from one climate to another.

One chimp immediately took center stage. Number 65, a 37-pound, 44-month-old male named Chang was, by far, the friskiest and the most enthusiastic chimp in the group. He was given a physical examination, checked on sensors, and tested by the psychomotor programmer. He passed all the tests with flying colors and was given the new name, HAM, an acronym for Holloman Aero Med where the chimp received his initial training.

The day of the sub orbital flight was set for January 31, 1961. The New Mexico Museum of Space History summarized the sequence of events leading up to take-off:

Ham the Astrochimp
(Photo courtesy of New Mexico Museum of Space History)

Nineteen hours before launch, both HAM and his back up were given a low-residue feeding, were fitted with biosensors, and checked in their pressurized couch-cabins. Seven and a half hours before lift-off, they were given a physical as well as several sensor and psychomotor tests. At sunrise, four hours before the flight, they were suited up, placed on their couches, and taken to the transfer van for the ride to the launch pad. At 7:53 a.m., Ham's couch-cabin was inserted into an MR-2 Mercury spacecraft which was atop an 83-foot-tall Redstone Rocket. At 11:55 a.m., the Mercury-Redstone lifted off the pad at Cape Canaveral.

During the 16-minute flight, HAM sustained about 17 Gs (17 times the force of gravity) during liftoff and reentry. He was weightless for about seven minutes. He was stable and worked the levers perfectly. The mission was a succes. However, Ham's capsule didn't land exactly where it was expected to in the Atlantic Ocean.

Since it would have taken the recovery ship, the destroyer Ellison, too long to reach Ham, a helicopter was dispatched from another ship, the Donner. When the helicopter reached Ham's capsule, the spacecraft was on its side and taking on water. It was estimated that about 800 pounds of water was in the spacecraft. Luckily, none was in Ham's compartment.

Once the capsule was taken aboard ship, it was opened. HAM smiled and eagerly accepted an apple and half an orange. He let his handlers hold him and he was his usual happy self. His mood changed only when his trainers tried to coax him back into his special "bio pack" for some photographs. He bared his teeth, apparently thinking they wanted him to taken another ride.

Following his 155 mile odyssey into space, HAM returned to Holloman Air Force Base where he spent the next two years "hamming it up" for newspapers and photographers (his picture even appeared on the cover of Life Magazine) and performing tasks to determine whether there were any harmful effects from his historic trip. But since he was healthy and there were no more flights in his future it was decided that he would retire.

In April 1963, HAM went to the National Zoological Park in Washington, D.C. He lived at the park for 17 years and was viewed by 50 million visitors. In 1980, following complaints by animal activists who worried that the chimp was suffering in his solitary enclosure, HAM was moved to a zoo in North Carolina where he became part of a social colony of chimpanzees. He then weighed 175 pounds and his beard was almost white but he made friends and found a special lady chimp to love.

HAM the Astrochimp died on January 17, 1983. His body was cremated and shipped to Alamogordo for burial at the International

Space Hall of Fame at the New Mexico Museum of Space History, a place dedicated to the memories of historic space missions, astronauts past and present, successes, tragedies, and even a chimpanzee once known as Number 65 that became America's first primate in space.

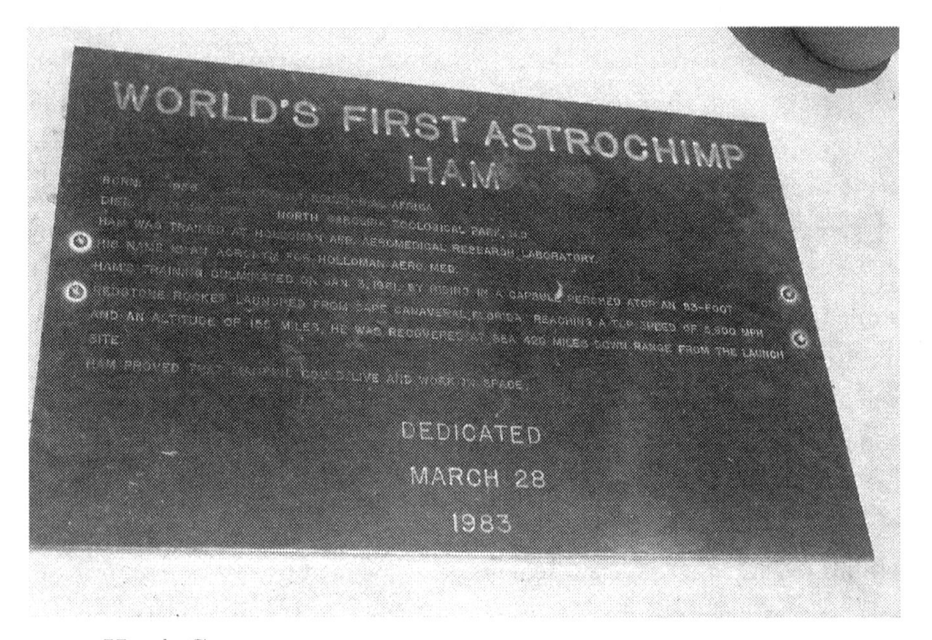

Ham's Grave

TRINITY SITE

Location: White Sands Missile Range, Highway 380, 12 miles east of San Antonio
Established: 1965
Features: Ground Zero Monument, McDonald Ranch
Contact: WSMR Public Affairs Office / (505) 678-1134

On Monday morning, July 16, 1945, the world was changed forever when the first atomic bomb was tested in an isolated area of the New Mexico desert. The nuclear blast created a flash of light brighter than a dozen suns. The light was seen over the entire state of New Mexico as well as in parts of Texas, Arizona, and Mexico. The resultant mushroom cloud rose to over 38,000 feet within minutes and the heat of the explosion was 10,000 times hotter than the surface of the sun! At ten miles away, this heat was described as like standing directly in front of a roaring fireplace. Every living thing within a mile of the blast was obliterated.

This is how the U.S. Department of Energy describes the first atomic bomb test, code named "Trinity," that took place on the Alamogordo Bombing and Gunnery Range, now known as the White Sands Missile Range, in the desolate Jornada del Muerto Valley.

The story of "Trinity" began 230 miles north at a one-time boys' ranch in Los Alamos, New Mexico, where scientists participating in the Manhattan Project led by Dr. J. Robert Oppenheimer gathered together in 1942 to design and build a bomb strong enough to bring about the end of World War II. The scientists came up with two designs—one using uranium 235 secured from Oak Ridge, Tennessee and another using plutonium from Hanford, Washington. Everyone was fairly confident that the uranium would work as desired but plutonium was a new element and no one was sure whether or not it would produce the essential chain reaction. A test was necessary.

Sign at Trinity Site

Eight locations were selected as possible test sites but based on its isolation, closeness to Los Alamos, and the fact that it was already under military control, the Jornada Valley near Alamogordo was chosen. Once used by Spanish explorers and colonists, the Jornada del Muerto (Journey of Death) was sixty miles of treacherous, waterless desert made even more dangerous because of frequent Indian attacks. Before the beginning of World War II, this sparsely populated area was mostly private grazing land but by January of 1942, the ranchers vacated their land and the War Department began using it as an artillery and bombing practice area. In 1944, part of the bombing range was set-aside as the Manhattan Project's "Trinity" test site. Observation points were set up, bunkers were built, a 100-foot steel tower was erected, and an old ranch house was transformed into an assembly station for the bomb's core.

On July 13, 1945, the plutonium was transported from Los Alamos to Trinity. After being carefully inserted into the core, it was taken to the 100-foot tower for insertion into the bomb mechanism. On the first try, the core wouldn't go in but after letting the temperatures of the plutonium and the casing equalize, the 15-pound core slid smoothly into place. The following morning, the completed "device" as it was called, was slowly raised to the top of the tower under which mattresses had been piled in case the bomb fell. Perched on top of the tower, scientists began installing the explosive detonators. Seismographic, photographic, and various other instruments for recording radioactivity, temperature, and air pressure were set up around the perimeter. Everything was ready.

A betting pool was started by scientists at Los Alamos on the possible yield of the Trinity test. The Nobel Prize-winning physicist Enrico Fermi was willing to bet anyone that the test would wipe out all life on Earth with special odds on the mere destruction of the entire State of New Mexico.

Although originally scheduled for 4 a.m. on July 16th, the test had to be delayed because of inclement weather. Forty-five-minutes later, the weather improved enough to allow the test to proceed. Countdown began at 5:10 a.m. The 19-kiloton bomb exploded at 5:29 a.m., and the rest, as they say, is history.

Because the site lies within the impact area for missiles fired into the northern part of the White Sands Missile Range, it is closed to the public except for the first Saturdays in April and October. After driving 17 miles past a security guarded gate, visitors walk through chain-link and barbed wire fences into an area where a black lava rock obelisk stands at Ground Zero, the spot where the 100-foot tower once stood. A short distance away, a replica of Fat Man, the bomb dropped on Nagasaki, sits on a flatbed trailer and beyond the trailer, a wooden shelter covers pieces of a green glassy substance that formed as the heat of the blast melted the surrounding desert sand. Known as Trinitite, the substance once covered the entire depression caused by the explosion. Years later, the depression was filled and much of the Trinitite was taken away by the Nuclear Energy Commission. Outside the fenced area, buses wait to take visitors to tour the reconstructed McDonald Ranch house where the core was assembled before being taken to the tower.

Trinity Monument

There's a lot of history in this place. Some people come here because they want to know more about war; others come because they want to know more about peace. Regardless of the reason, most people agree that:

All life on Earth has been touched by the event that took place here.

SANTA FE NEW MEXICAN

Los Alamos Secret Disclosed by Truman

ATOMIC BOMBS DROP ON JAPAN

Deadliest Weapons in World's History Made In Santa Fe Vicinity

CAPITOL

'Utter Destruction,' Promised in Potsdam Ultimatum, Unleashed; Power Equals 2,000 Superforts

May Be Tool To End Wars; New Era Seen

4 More Nippon Cities Now Smoldering Ruins

Hi Johnson Dies at 79

Tomato Juice Off Rationing

Now They Can Be Told Aloud, Those Stories of 'the Hill'

The Santa Fe New Mexican-August 6, 1945

6

GENEALOGY

GETTING STARTED IN GENEALOGY

The many cemeteries, monuments, and memorials in the state of New Mexico are dedicated to the memories of the men, women, and events that helped shape the history of the state. After visiting one (or more) of these sites, you may experience an urge to find out about more your ancestors and distant relatives. Who were they? Where did they come from? What contributions did they make to society? That's where genealogy comes in.

Although tracking down all your elusive progenitors might seem an interesting and worthwhile preoccupation, beware—it is time consuming, frustrating, and addictive. You will have to spend long hours in dusty old libraries or in front of a computer; you might have to forgo joining the bowling team; and family members may accuse you of losing touch with reality. However, if this is something you've made up your mind to do, the information in this chapter might be helpful.

To get started, you need to decide how far you want to go with your research. In other words, just how distant should distant relatives be? Do you just want to find out about direct line relatives (parents, grandparents, great-grandparents, etc.) or direct and collateral (siblings, aunts, uncles, cousins, etc.) relatives? A lot of people start out doing direct line research and end up collecting a lot of information about collateral relatives. Bear in mind the old theory about six degrees of separation—you might find out you're related to someone you would rather not even know. Either way, you're going to want to keep track of all the information you accumulate. The best way to do that is to write it down.

RESOURCES

Right about now, you're probably wondering where you're going to find all of this information. The logical answer is your relatives. Ask your mother, father, brothers, and sisters what they know or who they remember. Ask if any other relative has researched the family and if they are willing to share their information. Look through old family albums and see if you can identify all the people pictured in them. Whenever there's a birthday party or holiday celebration, take the albums with you. Can someone else identify the people you can't? If so, what do they know about them? Write to out-of-state relatives and ask them for information. You'd be surprised how willingly they open up once they find out you're working on a family tree. Plan a family reunion and invite all of your relatives—even the ones you haven't spoken to in years. Send out blank family record forms with the invitations and ask everyone to fill them out (as much as possible) and bring them to the reunion. When you finally run out of relatives, try someplace else.

________________ FAMILY RECORD FORM

Prepared by: __________________ **Date prepared:** ____________________________

HUSBAND:

	Date	City	State		
Born				Name of hospital:	
Christened				Name of church:	
Education				Name of HS/College	
Occupation					
Married				Name of church:	
Died				Cause:	
Buried				Name of cemetery:	
Notes:					

WIFE:

	Date	City	State		
Born				Name of hospital:	
Christened				Name of church:	
Education				Name of HS/College	
Occupation					
Married				Name of church:	
Died				Cause:	
Buried				Name of cemetery:	
Notes:					

CHILDREN:

Name	Date of birth	City	State	Name of spouse	Date of death
1					
2					
3					
4					
5					

OTHER INFORMATION

Visit or contact the hospitals, churches, and schools listed on the family record forms and find out how you can obtain copies of birth, christening, graduation, marriage, or death certificates for the relatives you have been able to identify. Sometimes these documents list the names of other relatives. If possible, send for birth, christening, graduation, marriage, and death certificates of those relatives as well.

Check out the county clerk's office and look through the land records. If a person's property was divided among his heirs, the records often show the heirs' names and addresses. Once you get that information, contact the heirs and ask them for their help.

Inspect the census records (www.census-online.com) The first census of the United States took place in 1790 and every ten years after that. Records dating from 1790 through 1880 are available for public inspection. The Federal census began in 1850 and includes not only the names of the head of the household but also the names of all the family members, their place of birth, their age, and their occupation. There's a wealth of information here if you're willing to spend the time to dig through it.

Look through old telephone directories. Try calling everyone with the surname you are researching. They might be long-forgotten relatives or, even better, they can lead you to someone else who is. Surf the web for surnames you've found. It may give you information about family members or lead you to other relatives doing similar research.

One of the best outside sources for family information is a library. Public libraries located in a relative's hometown have a lot of local information that can be used if you need background information and they might even have family history books, biographies, or newspaper articles about your relative. High schools and colleges usually keep old yearbooks that contain all sorts of information. The U.S. National Archives and Records Administration in Washington, D.C. (www.nara.gov) has old immigration lists and military records, the Library of Congress (www.lcweb.loc.gov/rr/genealogy) has publications from patriotic and hereditary societies, the Daughters of the American Revolution Library (www.dar.org/

library) and the Family Search Library of the Church of Jesus Christ of Latter-Day Saints (www.familysearch.org) may be able to provide material not available anywhere else.

RESEARCH HINTS

Always work from the present to the past—even though some patriot or politician has the same surname as yours, he is not necessarily related to you. Starting from you and working backwards is the only way you will be able to verify your hunches.

When doing research, be it in a local library or in aged Aunt's parlor, always check your references. Just because the source seems reputable, time sometimes warps reality. Town names change, buildings get torn down, old records get lost, distant cousins take on new names (and identities) and long-term memory gets muddled together with short-term. If you can't find the same information in at least three places—don't rely on it. But don't discard it either. It might come in handy somewhere down the line.

Be polite and respect the privacy of others. If you go barging in, asking all sorts of personal questions, you're liable to get the door slammed in your face. Remember—not everyone is willing to let the skeletons out of their closets.

Always include a self-addressed return envelope when sending family record forms to out-of-state relatives. If it doesn't cost them anything, they are more likely to return the form.

Record the title, author or publisher, date of publication, page number, library call number, and any other relevant information for every source you use. It will prevent you from reading the same information twice and it will make life a little easier in case you have to return to that source at a later date.

The more involved you become in genealogy, the more you will realize how little you really know. It's a convoluted, tricky business—check the web, buy a book, take a class, join a group. Some of the best websites are Ancestry.com (www.ancestry.com), RootsWeb (www.rootsweb.com),

U.S. GenWeb Project (www.usgenweb.com) and Robert Ragan's Pajama Genealogy site (www.amberskyline.com). As for books, the National Archives and Records Administration offers books on the Civil War, Black History, and various other research guides; Ancestry.com lists 50 books and two magazines; Genealogical.com lists 2000 books and CDs, and Amazon.com has more than 3000 listings including *Finding Your Roots* by Nancy Hendrickson, *Genealogy for Dummies* by Matthew and April Helm, *Your Guide to Cemetery Research* by Sharon Debartolo, and *Long Distance Genealogy* by Christine Crawford Oppenheimer. Local libraries, historic societies, and community colleges often offer seminars or classes on genealogy and they can usually put you in touch with one or more established genealogy groups such as the New Mexico Genealogical Society, the Hispanic Genealogical Research Center of New Mexico, or the New Mexico Jewish Historical Society. Whatever you do, learn as much as possible to insure you stay on the right track. You'll be glad you went the extra mile.

WHEN YOU THINK YOU'RE DONE

You've read the books, you've surfed the web, you've spent long hours in libraries, you've talked to all your relatives, you've documented your findings, and, finally, you've completed your family tree. Or have you?

Although it might be nice to think that, once you've documented your relatives all the way back to Adam and Eve, you're finished but, sorry to say, you are not. Several more things have to happen before you can say you are done.

For starters, make at least one copy of your research—all of it. Even though this sounds crazy, it's important. Things happen—files get lost, your computer crashes, someone spills coffee on your only copy, or a relative "borrows" your research and fails to return it. Having at least one extra copy insures that if the worst happens you won't have to start over again.

Take your research a step further and write biographies about each person. Add photos wherever possible and consider using the biographies as the basis for a book—either fiction or non-fiction. Even if you have to self-publish, wouldn't it be nice to have a book to pass on to family members and future generations.

Consider creating an old-fashioned "Family Tree" complete with trunk, branches, and twigs to give as a present to favorite relatives. Book stores, gift shops, and web sites offer several varieties or, if you are artistically talented, you can create your own.

Teach your kids how to carry on where you left off. Just like you, they might enjoy learning about their ancestors and, since you've already done a lot of the work, all they would have to do would be to add their generation. While we're on the subject of future generations, make sure you include your completed research and documentation in your will. Leave it to someone who will benefit from your work, even if it's only the public library.

Last, but not least, remember to hug your relatives

7

TOMBSTONE RUBBINGS

HISTORY OF TOMBSTONE RUBBING

Tombstone rubbing, sometimes referred to as frottage, dates back as far as the ancient Egyptians. When a person died and was entombed, the Egyptians carved stories, songs, poems, and prayers in relief style hieroglyphics on the stone work inside the tomb. Since the tomb was to be sealed, the only way that family members could keep a memento of the lovely hieroglyphics was to take a rubbing of it.

In the 17th and 18th centuries, people began leaving their European homelands to explore new territories. Never sure if they would ever return home, they often took rubbings of deceased loved ones tombstones. Sometimes it was all they had to show as physical evidence of the lives they once lived. As time progressed and people became even more mobile, they made rubbings of the family tombstones so that they didn't feel as if they were abandoning their loved ones. It was an inexpensive but effective way of ridding themselves of guilt.

In recent years, there has been a revival of interest in tombstone rubbing. Requiring minimum supplies and little or no experience, many people spend their summer vacations traveling all over the United States, searching overgrown cemeteries for abandoned gravesites, and taking rubbings of old tombstones. Many of the tombstones are of family members but some are of rich, famous, or notorious historic figures. In the process, rubbers get to spend a lot of time out-of-doors, learn something about their family background (and themselves), and meet a lot of interesting people. Once home, the rubbing is heated to "set it" and usually placed into a frame and displayed so that future generations can admire it.

TOMBSTONE SYMBOLISM

One of the first things most people wonder about is what all the ornate symbols and epitaphs on tombstones mean. Although many of the older symbols have lost their significance, some are still readily identifiable. For instance, lilies convey purity or innocence, roses immortal love, and poppies eternal sleep. The figure of a woman denotes sorrow or grief, an angel rebirth or resurrection, and the Grim Reaper the inevitability of death. Birds usually indicate eternal life, fish spiritual nourishment, and lions courage or bravery. A surprisingly large number of tombstone symbols represent family pets, sports cars, musical instruments, or bottles of beer. Evidently, even in death, many people found it hard to separate themselves from their favorite pastimes. As for epitaphs, most are straightforward but some reveal something about the person's character. Imagine the mindset of the person who wanted "I told you I was sick" or "Forgotten but not gone" written on his tombstone.

DO'S AND DON'TS

Regardless of the significance of their motifs or epitaphs, tombstones impart important information about the person buried beneath and must be treated with respect and reverence. Unfortunately, this is not always the case. In fact, some inexperienced or overzealous gravestone "rubbers" have caused so much damage that the Association for Gravestone Studies (www.gravestonestudies.org) issued a list of Do's and Don't s to safeguard tombstones from further damage.

Please DO

- Check with cemetery superintendent, cemetery commissioners, town clerk, historical society, or whoever is in charge to see if rubbing is allowed in the cemetery.
- Get permission and/or a permit as required.
- Rub only solid stones in good condition. Check for any cracks, evidence of previous breaks and adhesive repairs, defoliating stone with air pockets behind the face of the stone that will collapse under pressure of rubbing, etc.
- Become educated; learn how to rub responsibly.

- Use a soft brush and plain water to do any necessary stone cleaning.
- Make certain that your paper covers the entire face of the stone; secure with masking tape.
- Use the correct combination of paper and waxes or inks; avoid magic marker-type pens or other permanent color materials.
- Test paper and color before working on stone to be certain that no color bleeds through.
- Rub gently, carefully.
- Leave the stone in better condition than you found it.
- Take ALL trash with you and replace any gravesite materials that you may have disturbed.

Please DON'T

- Don't attempt to rub deteriorating marble or sandstone, or any unsound or weakened stone. For example, avoid any stone that sounds hollow when gently tapped or is flaking, splitting, blistered, cracked, or unstable on its base.
- Don't use detergents, soaps, vinegar, bleach, or any other cleaning solutions on the stone—no matter how mild!
- Don't use shaving cream, chalk, graphite, dirt, or other concoctions in an attempt to read worn inscriptions. Using a large mirror to direct bright sunlight diagonally across the face of a grave marker casts shadows in indentations and makes inscriptions more visible.
- Don't use stiff-bristled or wire brushes, putty knives, nail files, or any metal object to clean or to remove lichen from the stone. Soft, natural bristled brushes, whiskbrooms, or wooded sticks are usually OK if used gently and carefully.
- Don't attempt to remove stubborn lichen. Soft lichen may be thoroughly soaked with plain water and then loosened with a gum eraser or a wooden Popsicle stick. Be gentle! Stop if lichen does not come off easily.

- Don't use spray adhesives, scotch tape, or duct tape to attach paper to gravestone. Use ONLY masking tape.
- Don't use any rubbing method that you have not actually practiced under supervision.
- Don't leave masking tape, wastepaper, coloring objects, etc., at the gravesite.

If, after all these precautions, you've decided to proceed with a tombstone rubbing, the following information should be helpful.

HOW TO MAKE A TOMBSTONE RUBBING

First of all, gather your supplies together. You'll need at least one soft-bristled brush (no wire brushes, please), a role of masking tape, a pair of scissors or a retractable razor knife, newsprint, rice paper or interfacing fabric (butcher paper also works well), rubbing wax, crayon wedges or charcoal (don't use chalk or ink markers), a spray bottle filled with water (no detergent or chemicals), a large sponge, several old towels or rags, a pair of hand-held grass clippers, a cardboard tube or art portfolio, a roll of waxed paper, a kneeling pad, drinking water (aside from the water in the spray bottle), sunscreen, insect repellent, a first aid kit (just in case), a pencil and notepad, work gloves, and a hat. If you're going to be going to several different locations you might consider placing all of your supplies in a special "cemetery box" or basket so that you won't forget anything whenever you go out.

Get someone who has done several rubbings to show you how (or buy a book) and practice the techniques at home before leaving for the cemetery. Rocks, engraved patio stones, decorative clay pots, the living room lamps, or even old street signs make good practice models. Learn how to clean the surface of the object to be rubbed, how to attach the paper or fabric with as little tape as possible, and how to rub the image without tearing the paper or transferring any of the coloring medium to the surface. Practice, practice, practice but, and this cannot be stressed strongly enough: NEVER PRACTICE IN THE CEMETERY. The time to learn this art is before you leave home.

Once you think you're ready, call the cemetery to be sure they allow tombstone rubbings, get a permit if needed, determine a time when there will be no funerals being conducted, call a baby-sitter or send the kids off to school, pack a lunch, load up the car, and head on out.

When you arrive at the cemetery, choose a tombstone that is completely stable. Do not attempt a rubbing on any stone that is wobbly, crumbling, sounds hollow when tapped, or shows any signs of erosion or previous damage. Clip any weeds or grass that have grown up around the base of the stone, brush any accumulated dust from the surface of the stone, wash off any lichen if necessary, and make sure the surface is completely dry before proceeding—usually 20 to 30 minutes.

Cut a piece of paper or fabric about six inches larger than the tombstone, write any information about the stone on the back of the paper, and tape the paper to the stone making sure that the paper can't shift once you start rubbing.

Start from the outer edges and work toward the middle using long, even, strokes, all going in the same direction. Be gentle, work slowly and don't press too hard—you could rip the paper. Apply more pressure and continue to rub until the image achieves the darkness you desire. When the rubbing is done, carefully remove the paper from the tombstone, remove the tape from the paper, clean up any tape residue left on the stone, cover the completed rubbing with waxed paper and place it in a cardboard tube or art portfolio for safe transportation home. Pack up all of your tools and supplies, check the tombstone to be sure it hasn't been marked or damaged in any way (notify the cemetery if it has), pick up whatever grass you clipped or weeds you pulled, look around for scraps of paper or bits of trash, say a little prayer, and leave the site better than you found it.

PRESERVING A RUBBING

There are several different methods for preserving tombstone rubbings. Some people spray the paper or fabric with artist's fixative or hair spray to "set" the image and protect it from smudging. Others place the rubbing face-up on a paper-covered ironing board (protects the ironing board from image transfer), cover it with an old towel, and press down with a hot iron (never use a back and forth motion) to permanently "melt" the image into the paper or fabric

Once the rubbing is prepared, it may be attached to an acid-free artist's mat-board and placed in a glass-enclosed frame using spacers between the glass and the image to prevent damage from condensation or mildew. If the rubbing is to be placed in a book or album, make sure the image remains covered with waxed paper or acid-free tissue to avoid accidental transfer or smudging.

Properly preserved, most tombstone rubbings will last for several generations. They may be used as room decorations, illustrations for family trees, or patterns for craft projects. Some people produce tombstone rubbings and sell or donate them to historic societies, libraries, or out-of-state genealogical researchers.

If tombstone rubbing seems too harmful or abrasive, consider the alternatives: sketching, videotaping, or photography

PHOTOGRAPHING TOMBSTONES

Although many people recommend one type of camera over another, the truth of the matter is that any camera (including a disposable) will produce good quality, interesting photographs as long as the photographer knows how to use it. Even so, here are some tips that may increase your possibilities of success.

Experiment with different types of film (black & white, color, transparency, etc.) before taking any photos that cannot be re-shot at a future date. Consider different film speeds. Slower film (low numbers) requires more exposure to light but produces very detailed photographs while fast film (higher numbers) is good in low light. Tripods are great, especially if you'll be using low speed film or shooting from a low angle and a flash is important if you need to increase the amount of lighting hitting the face of the tombstone. Some photographers like to use skylight filters to reduce the blue tint in color photos taken above 5,000 feet, others prefer a color filter to control light intensities in black and white photos.

Choose the right time of day to photograph tombstones. Early morning and late afternoon are good because colors and shadows are more intense and there are fewer people around.

Check to see which way is the sun coming from. Does it reflect off the face of the tombstone or does it backlight it? If you need to increase the light, try using a mirror (full-length is best), aluminum foil, or an inexpensive space blanket.

Don't "clean" the tombstone unless it is absolutely necessary. More damage than good has been done by hobbyists trying to "clean up the area." Never use shaving cream, flour, peanut butter, mayonnaise, vinegar, butane, or anything other than a soft-bristled brush and water to clean away loose debris. As far as moss or lichens...why not just leave them where they are? They will add unexpected dimension to your photos.

The more pictures you take, the better. Try shooting from different angles to achieve unusual perspectives and lighting effects. Photograph surrounding tombstones to provide visual context. Eliminate any background elements that may be distracting. Make sure to take at least one picture where the inscription fills the entire camera frame. If the tombstone is leaning, tilt the camera to compensate for the tilt or just shoot the picture straight on to call attention to the tombstone's deterioration.

Anyone who has walked a cemetery knows that not all the tombstones are in great shape. In fact, some of them are in horrible shape. Some have been there for so long that moss has grown on them; others are sinking into the earth. Does that mean they are any the less valuable or that they are not worthy of being preserved? Of course not. Tombstones are our windows to the past. They tell us who our ancestors were, where they came from, and what they did while they were on this earth.

That's why we take photos and do tombstone rubbings and that, as you will learn in the following chapter, is why we try to preserve them.

8

PRESERVATION

PRESERVATION

New Mexico has a rich and diverse history that represents where we came from, where we have been, and what we have accomplished. Much of this history is embodied in the buildings, cemeteries, monuments, and cultural landscapes that surround us. For some people, these places are links to the past, for others they are guideposts to the future. Because New Mexico's history and traditions are among its greatest cultural and economic assets, and because their unique character makes them irreplaceable, they are worthy of protection. Yet, even as you read this book, many of these treasured resources are being lost or threatened because of urban renewal, pollution, vandalism, carelessness, neglect, or apathy. When historic buildings or monuments are torn down or cemeteries are allowed to deteriorate, an important part of our past disappears. Is anything being done to prevent that from happening?

On October 15, 1966, the United States Congress passed the National Historic Preservation Act (Public Law 89-665) that established the National Register of Historic Places and declared that:

1. *the spirit and direction of the Nation are founded upon and reflected in its historic heritage;*
2. *the historical and cultural foundations of the Nation should be preserved as a living part of our community life and development in order to give a sense of orientation to the American people;*
3. *historic properties significant to the Nation's heritage are being lost or substantially altered, often inadvertently, with increasing frequency;*
4. *the preservation of this irreplaceable heritage is in the public interest so that its vital legacy of cultural, educational, aesthetic, inspirational, economic, and energy benefits will be maintained and enriched for future generations of Americans;*
5. *in the face of ever-increasing extensions of urban centers, highways, and residential, commercial, and industrial developments, the present governmental and nongovernmental historic preservation programs*

and activities are inadequate to insure future generations a genuine opportunity to appreciate and enjoy the rich heritage of our Nation;

6. *the increased knowledge of our historic resources, the establishment of better means of identifying and administering them, and the encouragement of their preservation will improve the planning and execution of Federal and federally assisted projects and will assist economic growth and development; and*
7. *although the major burdens of historic preservation have been borne and major efforts initiated by private agencies and individuals, and both should continue to play a vital role, it is nevertheless necessary and appropriate for the Federal Government to accelerate its historic preservation programs and activities, to give maximum encouragement to agencies and individuals undertaking preservation by private means, and to assist State and local governments and the National Trust for Historic Preservation in the United States to expand and accelerate their historic preservation programs and activities.*

An underlying motivation in the passage of this Act was to transform the Federal Government from an agent of indifference, frequently responsible for needless loss of historic resources, to a facilitator, an agent of thoughtful change, and a responsible steward for future generations. In order to achieve this transformation, Section 101 of the Act included guidelines for the regulation of state historic preservation programs and the appointment of State Historical Preservation Officers (SHPO). The responsibilities of an SHPO were outlined to include, among other things, cooperation with federal and state governments, preparation and implementation of a statewide historic preservation plan, and communication with local governments to assist them in carrying out their historic preservation programs.

In 1969, the state of New Mexico assigned its first SHPO and passed the Cultural Properties Act "to provide for the preservation, protection, and enhancement of structures, sites, and objects of historical significance within the state, in a manner conforming with, but not limited by, the National Historic Preservation Act of 1966." In the following years several community-based volunteer organizations such as the New Mexico Heritage Preservation Alliance, the Historic Santa Fe Foundation, Cornerstones Community Partnerships, the Las Vegas Citizens Committee for Historic Preservation, the Tularosa Basin Historic Society, the Taos County Historical Society, and the City of Albuquerque Preservation Planning Program were formed to implement the objectives set forth by the National Historic Preservation Act and the New Mexico Cultural Properties Act. The goals of all of these organizations are identical: to safeguard the future of New Mexico's archaeological, cultural, and historic resources and to educate the public as to the value and importance of preserving them.

There are four separate levels of historic preservation—preservation, rehabilitation, restoration, and reconstruction. Preservation simply means repairing and maintaining a structure, as is, with all of the changes and alterations made to it over the years; rehabilitation means giving an under-utilized or abandoned building a new function or purpose; restoration means returning a property to its appearance and/or use during a certain time period; and reconstruction means building an authentic reproduction of a structure that no longer exists.

The first step in any historic preservation project involves conducting a survey to determine a site's historic significance. Is the site associated with a historic person or event? Does the architecture of the site reflect a distinctive style or construction method? Does the site exemplify the cultural, social, or historical heritage of a community? If the site is a cemetery, were any famous people buried there? Every aspect of the proposed project is then assessed and researched, photographs are taken, details are verified, funds or grants are applied for and, if appropriate, the site is nominated for addition to the National Registry of Historic Places. Once all the paperwork is complete, the project can begin.

Over the years, the combined efforts of various preservation groups throughout New Mexico have produced amazing results. Because of their can-do attitude, they have preserved old battlefields and trails; restored and revitalized numerous neighborhoods; saved abandoned schools from the wrecking ball and transformed them into productive enterprises; and restored neon signs along a road once filled with Model-T Fords and roadside motor courts.

In 2004, the New Mexico Heritage Preservation Alliance issued a list of the most endangered places in New Mexico. The list included Motel Boulevard in Lordsburg, the Aztec Ruins in San Juan County, Mesa Prieta in Rio Arriba County, the Hoyle House in White Oaks, the traditional village of Agua Fria in Santa Fe, Folsom Hotel in Union County, Lake Valley Ghost Town in Sierra County, the Valle Vidal Unit of the Carson National Forest, St. John's Methodist Episcopal church in Colfax County, and—possibly the most endangered—all the marked and unmarked cemeteries throughout New Mexico.

Not so long ago, it was the family, friends, and members of the community that performed all the activities associated with death. From building the coffin to digging the grave, death was dealt with on a personal basis. Tombstones were cleaned, fences were repaired, fresh flowers (plastic or real) were placed on the grave on a regular basis, and visitations were frequent. Long after they passed from this earth, the deceased were "remembered" through events such as Decoration Day and Dias de los Muertos—times when the family reunited to recall the past and honor their dead.

Since those days, things have changed. Today, with the transient nature of Americans and the lack of economic opportunities for the young in rural areas, many people have moved away from their ancestral homes. When they moved away, the care of their deceased loved ones fell to indifferent, hired workers. Using gas-powered weed whackers, industrial lawnmowers, and toxic chemicals, the workers trimmed, sprayed, and, all too frequently, caused damage around the otherwise ignored graves. Tombstones deteriorated, some succumbed to pollution, and old cemeteries took on an atmosphere of abandonment. Without any kind of regular surveillance, the graveyards became easy targets for vandals. Tombstones and monuments were knocked over, broken, vandalized, and even stolen. Graffiti was sprayed everywhere. Family treasures disappeared from gravesites, and in some instances, graves were robbed.

Saving Graves, a coalition of cemetery preservation advocates, states " Saving an endangered cemetery is not an easy project. It is not fun, and it is not something to be taken lightly. It requires a great deal of time, energy, and effort. The issues can become emotionally charged and can drag on for a considerable amount of time. (Advocates) are more likely to make enemies than to make friends as a result of (their) efforts." So why would anyone want to roll up his sleeves, spend his time and energy, and battle city hall just to become involved in a cemetery preservation project—or any other preservation project for that matter?

Historic preservation is both a public activity and a private passion. It adds value to the lives of New Mexican residents by teaching us about the events, people, and values that formed our communities, by creating an appreciation of local history, and by forming an indestructible connection to our past. It protects and saves our heritage by describing and documenting it. It fosters an appreciation for our diverse cultural heritage. It makes us an integral part of history by virtue of our protecting it. But it also stabilizes our neighborhoods, revitalizes downtowns, creates good jobs, attracts tourist dollars, and increases the tax base. Without historic preservation, places like the Santuario de Chimayo, Billy the Kid's Tombstone, D.H. Lawrence's Shrine, and the Mormon Battalion Monument would be lost forever.

Someone once said that a culture is identified by what it leaves behind. Where did its people come from? What did they believe in? What were their customs? How did they live? How did they die? Historic preservation answers these questions but it also educates and enriches the mind and spirit of the people by enhancing their quality of life and instilling a powerful sense of pride, permanence, and community. The cemeteries, monuments, and memorials left behind by our ancestors are tangible evidence of our past. They are our legacy and we must never forget that we are the stewards of that priceless legacy. It is our obligation to bequeath them, as unadulterated as possible, to future generations.

If you are interested in learning more about or becoming involved in the historic preservation of New Mexico, please contact one of the following:

New Mexico Heritage Preservation Alliance
(Santa Fe) 505-989-7745
Historic Santa Fe Foundation (Santa Fe)
Cornerstones Community Partnerships (Santa Fe)
505-982-9521 www.cstones.org
Los Alamos Historical Society (Los Alamos) 505-662-6272
Las Vegas Citizens Committee for Historic Preservation
(Las Vegas),
Taos County Historical Society (Taos)
City of Albuquerque Preservation Planning Program
(Albuquerque)
Tularosa Basin Historic Society (Alamogordo)
Doña Ana County Historical Society
(Las Cruces) 505-541-2155
Historical Society for Southeast New Mexico
(Roswell) 505-622-8333

It's our heritage—let's preserve it!

GLOSSARY

The rich Hispanic heritage of New Mexico is reflected in the names of many settlements, land formations, religious items, and everyday objects. Below is a small list of Spanish terms visitors may encounter. Knowing their meaning may lead to a better understanding and appreciation of New Mexico. Also included are several words that may not be familiar to all readers.

Acequia: Irrigation ditch
Alabados: Religious hymns
Atrio: Church courtyard
Anasazi: Prehistoric people of New Mexico and the 4 Corners Region
Angelitos: Deceased children
Bulto: Three-dimensional carved statue representing saint or holy person
Calaca: Humorous skeletal representation
Calavera: Sugar skull used during Day of the Dead
Calavario: Calvary or the hill where Penitentes perform symbolic crucifixion
Camposanto: Spanish cemetery
Capilla: Small chapel usually built for family use
Carreta de la muerte: Cart of death
Cedula Real: Royal decree
Cempazúchitl: Marigold-like yellow flowers used during Dias de los Muertos
Cenotaph: Monument erected for person (or persons) whose remains are buried elsewhere
Cerro: Hill or mountain

Chama: Spanish approximation for Tewa word meaning red

Columbarium: Chamber or wall in which urns containing the ashes of the dead are stored

Conquistador: Spanish conqueror

Corona: Floral offering

Cristo: Christ

Descanso: Roadside cross or resting place

Dias de los Muertos: Days of the Dead

Dicho: Old Spanish saying

Difunto: Deceased person

Doña Sebastiana: Powerful icon of death-Angel of death

El mas alla: The afterlife

El Vado: The ford (as in river)

Entierro: Burial

Flores enceradas: Crepe paper flowers dipped in wax

Frottage: Tombstone Rubbing

Genealogy: Research of family ancestry

Hueymiccailhuitontli: Aztec word for the adult dead

Iglesia: Church

La Semana Santa: Holy Week

Los Hermanos: A secular religious order that practiced acts of self-inflicted penance, also known as the Brotherhood or the Penitentes

Los Matines de las Tiniebas: Spy Wednesday (Holy Week) vespers

Low Rider: Unique automobiles with flamboyant graphics

Miccailhuitontil: Aztec word for the infant or young dead

Milagro: Miracle

Morada: Penitente meeting house and chapel

Museo: Museum

Nicho: Grotto-like structure used to enclose a Santo. Can be as short as 12 inches or tall enough to walk in

Nino: Boy child or Christ Child

Obelisk: Stone monument

Ofrenda: Multi-layered family altar containing flowers, pictures, candles, incense, and memorabilia

Pan de muerto: A sweet bread molded into the shape of a skull. Used during Dias de los Muertos

Papel picado: Cut paper banner used during Hispanic celebrations

Pasatiempo: The pastime

Pedernal: Flint

Penitentes: Short name for the Brotherhood of Los Hermanos

Pozito: Hole in the ground

Promesa: Obligation for fulfilled prayer

Pueblo: A village (Pueblos-native people who lived in villages)

Puerto del Sol: Gateway to the Sun

Rancho de las Golindrinas: Ranch of the Swallows

Reredo: Altar back or screen

Resador: Reader of holy books or prayers

Rio: River

Santa Cruz: Holy Cross

Santero: One who creates santos

Santo: Religious statue, altar screen or bulto representing a saint or holy person

Santo Entierro: Christ entombed

Santuario: Holy place

Semana Santa: Holy Week

Tierra bendita: Good Earth

Velorio: Wake

Vivos entre los Muertos: Living among the dead

BIBLIOGRAPHY

Portions of this book were based on information presented in the following:

Barkan, Rhoda and Peter Sinclaire. *From Santa Fe to O'Keefe Country: A One Day Journey Through the Soul of New Mexico*. Santa Fe: Ocean Tree Books, 1996.

Cash, Marie Romero. *Built of Earth and Song*. Santa Fe: Red Crane Books, 1993.

Chavez, Fray Angelico. *La Conquistadora: The Autobiography of an Ancient Saint*. Santa Fe: Sunstone Press, 1983.

Condie, Carol J. *The Cemeteries of Albuquerque, Bernalillo County, and Parts of Sandoval & Valencia Counties*. Albuquerque: Quivira Research Center, 1999.

Eisenstadt, Pauline. *Corrales: Portrait of a Changing Village*. Albuquerque: Cottonwood Printing Company, 1980.

Encinias, Miguel. *Two Lives for Oñate*. Albuquerque: University of New Mexico Press, 1997.

Fugate, Francis L. and Roberta B. *Roadside History of New Mexico*. Missoula: Mountain Press Publishing, 1989.

Garciagodoy, Juanita. *Digging the Days of the Dead*. Boulder: University Press of Colorado, 1998.

Golder, Frank Alfred. *The March of the Mormon Battalion from Council Bluffs to California*. New York: The Century Co., 1928.

Hogrefe, Jeffrey. *O'Keefe: The Life of an American Legend*. New York: Bantam Books, 1992.

Jaramillo, Cleofas M. *Shadows of the Past:Sombras del Pasado*. Santa Fe: Ancient City Press, 1972.

Jenkins, Myra Ellen and Albert H. Schroeder. *A Brief History of New Mexico*. Albuquerque: University of New Mexico Press, 1974.

Knaut, Andrew L. *The Pueblo Revolt of 1680: Conquest and Resistance in Seventeenth-Century New Mexico*. Norman: University of Oklahoma Press, 1964.

Lynn, Sandra D. *Windows on the Past: Historic Lodgings of New Mexico*. Albuquerque: University of New Mexico Press, 1999

Maurer, Stephen G. *Wild & Scenic Rio Chama*. Albuquerque: Public Lands Interpretive Association, 1999.

Nava, Margaret M. *Along the High Road: A Guide to the Scenic Route between Española and Taos*. Santa Fe: Sunstone Press, 2004.

Pettitt, Roland A. *Exploring the Jemez Country*. Los Alamos: Los Alamos Historical Society, 1990.

Prince, L. Bradford. *Spanish Mission Churches of New Mexico*. Glorieta: Rio Grande Press, 1915/1977.

Roberts, Brigham Henry. *The Mormon Battalion: Its History and Achievements*. Salt Lake City: The Deseret News Press, 1919.

Rudnick, Lois Palken. *Mabel Dodge Luhan: New Woman, New Worlds*. Albuquerque: University of New Mexico Press, 1984.

Sando, Joe S. *Pueblo Nations: Eight Centuries of Pueblo Indian History*. Santa Fe: Clear Light Publishers, 1992.

Simmons, Marc. *New Mexico: An Interpretive History*. Albuquerque: University of New Mexico Press, 1998.

Simmons, Marc. *The Last Conquistador: Juan de Oñate and the Settling of the Far Southwest*. Norman: University of Oklahoma Press, 1991

Steele, Thomas J. *Santos and Saints: The Religious Folk Art of Hispanic New Mexico*. Santa Fe: Ancient City Press, 1974.

Steele, Thomas J. *Works and Days: A History pf San Felipe Neri Church 1867-1895*. Albuquerque: Albuquerque Museum, 1983.

Snyder, Tom. *Route* 66 *Traveler's Guide and Roadside Companion* (*Second Edition*). New York: St. Martin's Griffin, 1995

Ungnade, Herbert E. *Guide to the New Mexico Mountains.* Albuquerque: University of New Mexico Press, 1965.

Wingard, Barbara & Jane Lester. *Telling Our Stories in Ways that Make Us Stronger.* Adelaide, Australia: Dulwich Centre Publications, 2001.

Wroth, William. *Images of Penance, Images of Mercy.* Norman: University of Oklahoma Press, 1991.